GASTRIC MOTILITY

GASTRIC MOTILITY
A Selectively Annotated Bibliography

Robert M. Stern
Pennsylvania State University

and

Christopher M. Davis
Simon Fraser University

Hutchinson Ross Publishing Company
Stroudsburg, Pennsylvania

Library of Congress Catalog Card Number: 82–12173
ISBN: 0–87933–430–4

84 83 82   1 2 3 4 5
Manufactured in the United States of America.

Library of Congress Cataloging in Publication Data
Stern, Robert Morris, 1937–
Gastric motility.
Includes bibliographical references and indexes.
1. Stomach—Motility. I. Davis, Christopher M.
II. Title. [DNLM: 1. Gastrointestinal motility—
Abstracts. ZWI 102 S839g]
QP151.S73 1982   016.612'32   82–12173
ISBN 0–87933–430–4

Distributed worldwide by Van Nostrand Reinhold Company Inc.,
135 W. 50th Street, New York, NY 10020.

PREFACE

The authors were both introduced to the topic of gastric motility by R. C. Davis at Indiana University in the late 1950s and early 1960s. In our separate laboratories at The Pennsylvania State Uiversity (Stern) and Simon Fraser University (Davis), we have continued to study various measurement techniques, particularly surface recording with cutaneous electrodes—the electrogastrogram or EGG. As psychophysiologists we began our search of the literature looking for studies that attempted to relate some aspect of behavior to gastric motility and quickly ascertained that few such studies have been published. Consequently, the search was broadened to include all studies involving gastric motility.

We started work on this bibliography in 1976 and it has taken six years to complete the project not only because we have had other obligations during this period but because in many cases the references were not easy to find. Since the study of gastric motility has not been in vogue for the past 50 years, it was a rare pleasure for us to be able to obtain several new citations from the reference pages of a textbook or review article. We also observed that, as in many other disciplines, individuals in this field who write in English tend to ignore the foreign literature. Thus a considerable effort was made to include in this bibliography numerous foreign references (e.g., Russian, Scandinavian, French, Japanese, and German).

We have found several articles relevant to our own research in the course of preparing this bibliography. We hope that our effort will stimulate other investigators and, perhaps, keep them from reinventing the wheel.

We would like to thank the interlibrary loan personnel in the Penn State Library for their assistance in obtaining many books, articles, and abstracts for us. The majority of the items were lent by the libraries of the Hershey Medical Center and the College of Physicians and Surgeons in Philadelphia, for which we are grateful. Many of the German articles were obtained from the library of the University of Mainz. We also acknowledge the help of Jonathan Cofer, Takami Watanabe, and Elizabeth Moes for translations; Thomas R. Bonifield for the computer searches of the literature; and Esther Strause, Ginger Swanger, Debbie Snyder, Joy Creeger, and Barb Robinson for typing. The indexes and computer printout of the manuscript were prepared by Kirk Evans. The authors are also grateful to the following students who spent many hours in the library uncovering relevant items that either predate or were missing from the computer-assisted searches: Nancy Denari, Stephen Macciocchi, Brad Perry, and Howard Crawford.

This project was partially supported by a grant from the Central Fund for Research, Pennsylvania State University.

Robert M. Stern
Christopher M. Davis

GASTRIC MOTILITY

INTRODUCTION

Empirical observation of human stomach motility began in earnest in 1825 with William Beaumont's observations of his patient, Alexis St. Martin, who had suffered an accidental gunshot wound that resulted in a permanent gastric fistula (Beaumont, 1833). These observations and later experiments on fistulous patients and animals conducted primarily by German and Russian physiologists provided the bulk of the early information on gastric motility in digestion and disease. Extensive and important as they were, the observations tended to be qualitative rather than quantitative, and since they necessarily came from a very limited, and always abnormal population, generalizations were difficult to make.

With the discovery of X-rays in 1895, Walter B. Cannon (1898) and, slightly later, Walter C. Alvarez (1968) began to study the movements of the stomach. At first a series of individual X-rays were used to make motion pictures. This procedure was eventually replaced by films (and today, video tapes) made directly from a fluoroscope screen. To be sure, motility can be examined by fluoroscopy for a limited period of time, although with this method quantification is also difficult, conditions are abnormal, and a definite risk to health exists.

During the first half of this century the most common method of studying gastric motility used pressure recording systems attached to swallowed balloons. Many years passed before investigators realized that the inflated balloons stimulated the stomach and, indeed, brought about contractile activity. More recently, direct measures of mechanical and electrical activity associated with motility involve placing transducers or electrodes in or on the stomach and, therefore, share with the balloon method the double disadvantage of affecting the measured response and restricting the situations in which stomach activity can be examined. The same disadvantages arise, though possibly to a lesser degree, with swallowed metal balls, open-tipped catheters, tiny swallowed transducers (endoradiosones) that transmit pressure, and even fiber optic endoscopes.

It is not surprising, then, that research on the control of stomach motility has come largely from in vitro and in vivo electrophysiological studies of animal preparations. Precontractile and contractile events in segments of guinea pig tenia coli and rat jejunum provided the data upon which models of the control of human motility were built. However, recent advances in techniques of biomedical measurement have opened the way for research on digestive and nondigestive motility in the undisturbed human gastrointestinal (g.i.) tract. We refer here to electrogastrography (EGG), the recording of gastric potentials from the surface of the skin. We will say more about this recording technique after briefly discussing control and innervation of the g.i. tract.

The Control of Motility

If a small piece of smooth muscle tissue from the g.i. tract of any of several species is deprived of its extrinsic nerve supply and placed in a nutrient solution it will likely commence contracting and relaxing rhythmically and continue to do so for many hours. If contractions are absent, or inhibited by atropine, intra- or extra-cellular recording reveals a rhythmic waxing and waning of electrical potential, a slow wave, of the same frequency as the contractions (when they do occur). This slow wave is so characteristic of smooth muscle tissue throughout the g.i. tract, except for the esophagus and upper stomach—in vivo as well as in vitro—that it has been termed the "basic electrical rhythm." This basic electric rhythm, also called electrical control activity (Sarna, 1975), turns out to be the necessary condition for contractions of g.i. smooth muscle and controls the timing of contractions as well.

Control activity originates in the longitudinal smooth muscle layer and spreads to the circular layer electrotonically. The period of control activity varies with tissue site, organism, and neural and hormonal influences. No intraluminal pressure changes are associated with control activity; it occurs continuously in the absense of mechanical contraction.

Control activity makes the membrane temporarily susceptible to the generation of a spike or second potential. When it occurs, the second potential, or electrical response activity (Sarna, 1975), consists of one or several spikes superimposed on the slow terminal-decline phase of the control activity and coincides with contraction. Response activity continues for about 12 seconds and always follows control activity within 4–8 seconds. "It appears that the initial potential initiates in the membrane some change essential for the production of contraction, and hence phases and coordinates contractile activity" (Daniel and Irwin, 1968, p. 1976).

In order for mixing and moving of the contents of the g.i. tract to occur, coordination of adjacent smooth muscle units both around the circumference of a segment and longitudinally between functional segments would be required. In intact organisms control activity occurs simultaneously around the circumference of a segment and spreads electrotonically to adjacent segments, providing the stimulus for control activity in them. The linkage between segments is apparently not neural; control activity artificially created in the body of the stomach produces control activity both proximally and distally, contrary to normal synaptic behavior and at an extremely slow rate. Moreover, atropine present at the membrane has no effect on control activity.

Though it appears in vivo that a wave of depolarization moves along the organ in a manner similar to the conduction of a nerve impulse, it is more accurate to picture the smooth muscle as composed of a series of independent oscillators that can be driven by other oscillators. Just such a chain of loosely coupled relaxation oscillators has been proposed by both Nelsen (1971) and Sarna et al. (1972) to explain the coupling and uncoupling of successive smooth muscle segments. Besides providing a physically defined descriptive terminology—that is, period, phase lag, and phase lock—the appropriate equations are able to predict the uncoupling of segments, variations in frequency, and the effects of transection of the bowel, among other events.

In vivo control activity usually originates proximally, successively driving more distal segments (since control activity renders the membrane insensitive to immediate reexcitation the spread is undirectional). In the human stomach, for instance, control activity ordinarily originates in the upper body in one or more cell groups. No histologically distinct pacemaker cells have been found, but the chain oscillator model predicts that those cell groups with the highest frequency would, in effect, become pacemakers since they would drive adjacent segments. No regular control waves have been observed in the fundus. The rhythm of gastric control activity is 3/min in the fasting human subject. It is this activity that determines the frequency of response activity and contractions, when they occur.

It is in the linkage between control activity and response activity and somewhat in the period of the control activity that extrinsic neural and hormonal effects are seen. Contraction may follow every control wave, be entirely absent, or follow some proportion of them. The presence of acetylcholine at the cell membrane greatly increases the linkage, but only if it arrives during the terminal phase of control activity. Catacholamines, epinephrine, and norepinephrine, also reduce or abolish the response activity when arterially administered (Daniel and Irwin, 1968), suggesting one mechanism for neural and hormonal inhibition of gastric motility.

Innervation of the Gastrointestinal Tract

The intrinsic nervous innervation of the g.i. tract is composed of local afferent fibers terminating on cell bodies in the myenteric and submucous plexuses within the organ. Axones of these cells terminate in other plexuses or return to supply smooth muscle cells of the same or nearby segments. However, motoneurones of the intrinsic system do not affect the smooth muscle fibers in a manner similar to that in striate muscles, where each muscle fiber has an associated axon branch serving it through a motor end plate. Rather, neurones in the g.i. tract run along, but not in contact with, several muscle fibers. Along their terminal length there are vericosities containing granular vesicles thought to contain transmitter substance or precursor so that neural effects are realized by a group of smooth muscle cells and result from a buildup of transmitter in the intercellular space. According to Kosterlitz (1968): "The motility of the stomach and intestines is controlled by intrinsic reflexes that are modulated by impulses reaching the intramural plexuses from extrinsic motor and inhibitor nerves" (p. 2147).

Kosterlitz points out that, in general, section of the entire extrinsic nerve supply to the stomach results in weakened peristalsis, diminution of tone, and delayed emptying.

Central neural activity and other bodily events must, of course, influence g.i. motility through some patterns of vagal and splanchnic activity, including effects on the intrinsic innervation and the release of adrenal catacholamines. Afferent fibers arising within the g.i. tract are involved in intrinsic and spinal reflexes via spinal afferents, and splanchnic efferents and brain stem reflexes via vagal afferents.

Not surprisingly, the behavior of the intact g.i. system is not well understood. However, the main function of the tract is digestion and evacuation of the nondigestable and of the by-products of the digestive process. Motile activity successively mixes and moves food along a path of 4–5 meters, a journey that takes from 12 hours to about 7 days to complete.

Surface Recording of Abdominal Potentials

It is likely that some of the electrical potentials generated within the tract could be recorded wth the aid of cutaneous electrodes attached to the skin over the abdomen—the familiar problem of recording from a distance, potentials occurring in a volume conductor. In the 1950s, R. C. Davis rediscovered the technique that was first described by Alvarez (1922) and Tumpeer and Phillips (1932) for recording g.i. potentials from the surface of the skin. Davis published two journal articles (Davis et al., 1957, 1959) and numerous technical reports dealing with the validity of the electrogastrogram (EGG). The research of Davis and other psychologists with the EGG is summarized in Russell and Stern (1967). More recently, medical researchers in this country (e.g., Nelsen and Kohatsu, 1968), France (e.g., Martin and Thillier, 1971), and Russia (e.g., Androsov et al., 1973) published EGG studies ranging from the very basic—relating the gastric electrical control activity and electrical response activity to the EGG—to the applied—differentiating the EGG pattern of a patient with stomach cancer from that of a patient with no gastric pathology.

Psychophysiologists are currently showing a renewed interest in both the interpretation and application of the EGG. Whereas Alvarez, Tumpeer, and Phillips, and Davis and his co-workers assumed the the EGG was a potential resulting from the contraction of the smooth muscle of the stomach, Nelsen and Kohatsu pointed out that the interpretation of the surface-recorded 3/min signal (the EGG) is more complicated. According to Nelsen and Kohatsu, the EGG is a measure of the electrical control activity (ECA), the ever present waxing and waning of the potential that occurs three times per minute in the human stomach. According to these same authors, the EGG does not inform us of the presence or absence of contractions, but when contractions do occur they are time-locked to the ECA. Most researchers working in this area agree that the frequency of the EGG is a valid measure of the frequency of ECA and contractions, when they occur. There is disagreement, however, over the relationship of the motor activity of the antrum and the interpretation of amplitude changes in the EGG. Brown et al. (1975) found that following a meal the EGG increased in amplitude by 150 percent, but the authors attributed the increase to the increased proxity of the stomach muscle to the surface electrode. In a more recent paper, Smout et al. (1980) point out that in their study, EGG was greater in amplitude when ECA and electrical response activity (ERA) (and contractions) were present than during ECA alone. And this difference was evident even when their subjects' stomachs were empty, suggesting that the increased amplitude of the EGG during ERA and contractions was not an artifact of stomach distention.

Martin and Thillier (1971) and Rebrov (1974) reported that the EGG could be recorded from the limbs as well as with cutaneous electrodes over the abdomen. In fact, Rebrov pointed out, there are advantages to recording from the limbs: less respiration artifact, less intestinal potential, easier accessibility to the electrode location, and so on. Rebrov found that hypermotility, when evident in a subject, was reflected in recordings from the surface of the abdomen as well as from the wrists and ankles. Stern and Stacher (1982) have demonstrated that the EGG recorded from the limbs increases in amplitude with eating as does the EGG from the abdomen.

Recent applications of the EGG have been reported in this country by Walker and her colleagues and by Hölzl and his colleagues in Germany. Walker and Sandman (1977) found that one aspect of the EGG differentiated among ulcer patients, arthritics, and healthy individuals. In a second study, Walker et al. (1978) reported that some individuals can learn to modify the tonic component of the EGG. Hölzl et al. (1979) described a new method of measurement and analysis of gastric motility, conjoint spectral gastrography, that combines EGG and magnetogastrography with spectral analysis. The authors report their use of this measure to assess the effects of stress on gastric motility. A summary of the procedure used to record the EGG can be found in Stern et al. (1980).

In the Bibliography a summary is provided for those articles in which the use of the EGG is described, thus reflecting the authors' interest in this measure, the only index of gastric motility that does not involve putting something inside the subject's stomach, is completely safe, and yields long, continuous records.

The titles of articles in recent gastrointestinal journals and papers presented at professional meetings and edited books indicate that there is an increase in interest in gastric motility. Books edited by the following individuals have been the result of recent meetings of the International Symposium on Gastrointestinal Motility: Vantrappen (1976), Duthie (1978), and Christensen (1980). (A new association, the American Motility Society, held its second biennial meeting in 1982.) Hölzl and Whitehead (1983) have just edited a new volume, entitled *The Psychophysiology of the Gastrointestinal Tract.*

References

Alvarez, W. C., 1922, The Electrogastrogram and What It Shows, *Am. Med. Assoc. J.* **78**:1116–1119.

Alvarez, W. C., 1968, Early Studies of the Movements of the Stomach and Bowel, in *Handbook of Physiology, Section 6: Alimentary Canal,* vol. IV: *Motility,* C. F. Code, ed., American Physiological Society, Washington, D.C., pp. 1573–1578.

Androsov, N. S., P. P. Firsova, and P. I. Tolstykh, 1973, Significance of the Electrogastrogram in Evaluating Gastric Motor Function in Cancer, *Vopr. Onkol.* **19**:39–43.

Beaumont, W., 1833, *Experiments and Observations on the Gastric Juice and the Physiology of Digestion,* F. P. Allen, Plattsburgh, 279p.

Brown, B. H., R. Smallwood, H. I. Duthie, and C. J. Stoddard, 1975, Intestinal Smooth Muscle Electrical Potentials Recorded from Surface Electrodes, *Med. Biol. Eng.* **13**:97–103.

Cannon, W. E., 1898, The Movements of the Stomach Studied by Means of the Roentgen Rays, *Am. J. Physiol.* **1**:13–14.

Christensen, J., ed., 1980, *Gastrointestinal Motility,* Raven, New York, 513p.

Daniel, E. E., and J. Irwin, 1968, Electric Activity of Gastric Musculature, in *Handbook of Physiology, Section 6: Alimentary Canal,* vol. IV: *Motility,* C. F. Code, ed., American Physiological Society, Washington, D.C., pp. 1969–1984.

Davis, R. C., L. Garafolo, and F. P. Gault, 1957, An Exploration of Abdominal Potentials, *J. Comp. Physiol. Psychol.* **50**:519–523.

Davis, R. C., L. Garafolo, and K, Kveim, 1959, Conditions Associated with Gastrointestinal Activity, *J. Comp. Physiol. Psychol.* **52**:466–475.

Duthie, H. L., ed., 1978, *Gastrointestinal Motility in Health and Disease,* MTP Press, Lancaster, England, 684p.

Hölzl, R., and W. Whitehead, 1983, *The Psychophysiology of the Gastrointestinal Tract,* Plenum, New York.

Hölzl, R., G. Schröder, and H. Kiefer, 1979, Indirect Gastrointestinal Motility Measurement for Use in Experimental Psychosomatics: A New Method and Some Data, *Behav. Anal. Modif.* **3**:77–97.

Kosterlitz, H. W., 1968, Intrinsic and Extrinsic Nervous Control of Motility of the Stomach and the Intestines, in *Handbook of Physiology, Section 6: Alimentary Canal,* vol. IV: *Motility,* American Physiological Society, Washington, D.C., pp. 2147–2172.

Martin, A., and J. L. Thillier, 1971, Electrogastroenterography (E.G.E.G.), *Presse Med.* **79**:1235–1237.

Nelsen, T. S., 1971, A Theory of Integrated Gastrointestinal Motor Activity Based on the Chain Oscillator Model, *Am. J. Dig. Dis.* **16**:543–547.

Nelsen, T. S., and S. Kohatsu, 1968, Clinical Electrogastrography and Its Relationship to Gastric Surgery, *Am. J. Surg.* **116**:215–222.

Rebrov, V. G., 1974, Diagnostic Possibilities of Electrogastrography from Parts of the Body Distant from the Stomach, *Klin. Med.* **52**:115–120.

Russell, R. W., and R. M. Stern, 1967, Gastric Motility: The Electrogastrogram, in *Manual of Psychophysiological Methods,* P. H. Venables and I. Martin, eds., North Holland, Amsterdam, pp. 219–244.

Sarna, S. K., 1975, Gastrointestinal Electrical Activity: Terminology, *Gastroenterology* **68**:1631–1635.

Sarna, S. K., E. E. Daniel, and Y. J. Kingma, 1972, Effects of Partial Cuts on Gastric Electrical Control Activity and Its Computer Model, *Am. J. Physiol.* **223**:332–340.

Smout, A. J. P. M., E. J. van der Schee, and J. L. Grashuis, 1980, What Is Measured in Electrogastrography? *Dig. Dis. Sci.* **25**:179–187.

Stern, R. M., and G. Stacher, 1982, Recording the Electrogastrogram from Parts of the Body Surface Distant from the Stomach (Abstract), *Psychophysiology* **19**:350–351.

Stern, R. M., W. J. Ray, and C. M. Davis, 1980, *Psychophysiological Recording,* Oxford University Press, New York, 256p.

Tumpeer, K. H., and B. Phillips, 1932, Hyperperistaltic Electrographic Effects, *Am. J. Med. Sci.* **184**:831–836.

Vantrappen, G., ed., 1976, *Proceedings of the Fifth International Symposium on Gastrointestinal Motility,* Typoff Press, Herentals, Belgium.

Walker, B. B., and C. A. Sandman, 1977, Physiological Response Patterns in Ulcer Patients: Phasic and Tonic Components of the Electrogastrogram, *Psychophysiology* **14**:393–400.

Walker, B. B., C. A. Lawton, and C. A. Sandman, 1978, Voluntary Control of Electrogastric Activity, *Psychosom. Med.* **40**:616–619.

BIBLIOGRAPHY

It is the authors' hope that this bibliography will provide the interested reader with easy access to the literature on gastric motility in general and surface recording in particular. We included all the references we found on stomach motility of terrestrial and amphibious vertebrates and references complete with summaries of those articles in which surface recording was used. Where available, the summaries are from the original authors' abstracts.

After an alphabetical listing of each reference by first author and summaries where applicable, the reader will find an Author Index, a Subject Index, and a Species Index. A given reference may be found under more than one category in the indexes. The literature was searched through 1975. Two computer-assisted searches were conducted, one using the MEDLAR (Medical Literature Analysis and Retrieval) system and the other NCMHI (National Clearing House of Mental Health Information) at the U.S. Public Health Service, Alcohol, Drug Abuse and Mental Health Administration. Descriptors used included gastrointestinal, gastric and stomach; and motility, contractions, waves, and potentials. A purposely overlapping search of the literature was made by the authors using bibliographic references such as *Index Medicus, Gastroenterology Abstracts,* and *Psychological Abstracts,* all of which should have been included in the data base for the computer searches. References from appropriate chapters in *Annual Review of Physiology* (Annual Reviews, Inc.) and the *Handbook of Physiology* (Williams & Wilkins Co.) were included as well as references obtained from textbooks and review articles in the field of gastroenterology.

1 – 12

1. Abeshouse, S. Gastro-intestinal manifestations accompanying diseases in the upper urinary tract. Am. J. Dig. Dis., 1935, 2, 477-494.

2. Abo, M. Photo-optical measurement of gastric motility--the relation between gastric type and its motility. Jap. J. Smooth Muscle Res., 1971, 7, 213-223.

3. Abrahamsson, H. Studies on the inhibitory nervous control of gastric motility. Acta Physiol. Scand., 1973, 89, 1-38.

4. Abrahamsson, H. Vagal relaxation of the stomach induced from the gastric antrum. Acta Physiol. Scand., 1973, 89, 406-414.

5. Abrahamsson, H. Reflex adrenergic inhibition of gastric motiliy elicited from the gastric antrum. Acta Physiol. Scand., 1974, 90, 14-24.

6. Abrahamsson, H. & Jansson, G. Elicitation of reflex vagal relaxation of the stomach from pharynx and esophagus in the cat. Acta Physiol. Scand., 1969, 77, 172-178.

7. Abrahamsson, H. & Jansson, G. Vago-vagal gastro-gastric relaxation in the cat. Acta Physiol. Scand., 1973, 88, 289-295.

8. Abrahamsson, H., Jansson, G. & Martinson, J. Vagal relaxation of the stomach induced by amorphine in the cat. Acta Physiol. Scand., 1973, 88, 296-302.

9. Abrahamsson, H., Johnsberg, T., Jansson G. & Martinson, J. Control of gastric receptive relaxation before and after vagotomy. In E. E. Daniel et al. (Eds.) Proc. 4th International Symposium on Gastrointestinal Motility. Vancouver: Mitchell Press, 1974.

10. Abrahamsson, H. & Thoren, P. Vomiting and reflex vagal relaxation of the stomach elicited from heart receptors in the cat. Acta Physiol. Scand., 1973, 88, 433-439.

11. Abrams, G. D. & Bishop, J. E. Effects of the normal microbial flora on gastrointestinal motility. Proc. Soc. Exp. Biol. Med., 1967, 126, 301-304.

12. Adair, G. S. & Goodman, E. N. A note on the potential difference across the stomach membranes in human subjects and a simple type of calomel electrode. J. Physiol., London, 1936, 87, 35.

13. Aeberhard, P. & Bedi, B. S. Proceedings: Effect of proximal gastric vagotomy and total vagotomy on the co-ordination of electrical activity of the stomach and duodenum. Br. J. Surg., 1975, 62, 153.

14. Agadzhanian, G. I. & Torosian, A. A. Change in gastric motor function in chronic gastritis under the influence of treatment with Dilizhan mineral water. Ter. Arkh., 1972, 44, 57-61.

15. Agostoni, E., Chinnock, J. E., Daly, M. & Murray, J. G. Functional and histological studies of the vagus nerve and its branches to the heart, lungs and abdominal viscera in the cat. J. Physiol., London, 1957, 135, 182-205.

16. Aguggini, G., Cheli, R., Maffeo, G., Mesen, A. P., Quarenghi, F. & Secchi, C. Preliminary observations on abomasal motility in calfs. Arch. Vet. Ital., 1968, 19, 251-256.

17. Alekseenko, N. Iu. & Voronin, L. G. Movements of the empty stomach of the rhesus monkey. Bull. Acad. Sci. USSR Ser. Biol., 1944, 3, 177-184.

18. Alexander, F. The influence of psychologic factors upon gastrointestinal disturbances. General principles, objectives and preliminary results. Psychoanal. Quart., 1934, 3, 501.

19. Alexander, F. Experiments on the horse stomach. Quart. J. Exp. Physiol., 1951, 36, 139-147.

20. Alexander, F. & Benzie, D. A radiological study of the digestive tract of the foal. Quart. J. Exp. Physiol., 1951, 36, 213-217.

21. Alexander-Williams, J., Donovan, I. A., Gunn, I. F., Brown, A. & Harding, L. K. The effect of vagotomy on gastric emptying. Proc. Royal Soc. Med., 1973, 66, 1102-1103.

22. Ali, T. M. & Singleton, A. G. Effect of immobilon on the oesophageal and stomach movements in sheep. Vet. Rec., 1974, 94, 172-173.

23. Allison, P. R. Function and dysfunction at cardia. Bull. Johns Hopkins Hosp., 1956, 99, 182-189.

24. Alvarez, W. C. Differences in rhythmicity and tone in different parts of the wall of the stomach. Am. J. Physiol., 1916, 40, 585-602.

25. Alvarez, W. C. Recent advances in gastric physiology. Am. J. Med. Sci., 1919, 158, 609-618.

26. Alvarez, W. C. New methods of studying gastric peristalsis. *J. Am. Med. Assoc.*, 1922, *79*, 1281-1284.

Knowledge of gastric physiology is still largely in the crude stage of purely visual observation. The problem is now being attacked with methods for obtaining multiple and simultaneous mechanical and electrical records from the wall of the stomach. Something has been learned about the gastric pacemaker; about different types of rhythmic activities; about gastric "systoles"; about "blocks" and dissociations of activity between the body of the stomach and the pyloric antrum. Stimulation of the vagus produces quite different effects in the fundus and in the antrum. It is surprising to find areas on the stomach in which the muscle is responding at one time to two or more sets of rhythmic impulses. Some observations have been made on the influence of the stomach on the duodenum. The records generally show little sign of the "law of the intestine". The fundus sometimes appears to relax while the rest of the stomach is contracting. The large gaps which exist in our knowledge must be filled if the surgeon is to operte safely on diseased stomachs.

27. Alvarez, W. C. The electrogastrogram and what it shows. *J. Am. Med. Assoc.*, 1922, *78*, 1116-1119.

The first steps have been taken in working out the technic of electrogastrography. The records shows that many of the waves which seem to begin in the lower third of the stomach have come as shallow ripples from a pace-making region near the cardia. There are, however, a number of different types of gastric peristalsis, and the stomach can change suddenly from one to another. At times, there seems to be a dissociation between the activities of the fundus and the pars pylorica. The fundus can contract several times to the antrum's once. Small, frequent waves are sometimes superimposed upon the larger, slower ones. These findings have been confirmed with mechanical recording devices. Marked blockage of the waves has been observed often in the preantral region. The conduction time of waves traveling down the stomach shows wide variation. Marked tonus changes occur in the duodenum coincident with the arrival of gastric waves at the pylorus. This observation may throw light on the production of pain in duodenal ulcer. The peristaltic rushes down the bowel are shown to originate in gastric waves. The first human electrogastrograms are here presented.

28. Alvarez, W. C. The Mechanisms of the Digestive Tract. New York: Hoeber, 1922.

29. Alvarez, W. C. A nonpolarizable electrode clamp designed for gastrointestinal work. Am. J. Physiol., 1924, 69, 249-253.

30. Alvarez, W. C. The Mechanics of the Digestive Tract (2nd Ed.). New York: Hoeber, 1928.

31. Alvarez, W. C. Physiologic studies on the motor activities of the stomach and the bowel in man. Am. J. Physiol., 1929, 88, 650-662.

32. Alvarez, W. C. Applied physiology of the stomach and duodenum. In G. B. Eusterman & D. C. Falfour (Eds.), The Stomach and Duodenum. Philadelphia: Saunders, 1935.

33. Alvarez, W. C. Congenital absence of cardia, case report. Gastroenterology, 1945, 5, 135-136.

34. Alvarez, W. C. An Introduction to Gastroenterology (4th Ed.). New York: Hoeber, 1948.

35. Alvarez, W. C. Early Studies of the movements of the stomach and bowel. In C. F. Code (Ed.), Handbook of Physiology. Sect. 6, Alimentary Canal. Vol. IV. Motility. Washington, D.C.: American Physiological Society, 1968.

36. Alvarez, W. C. & Mahoney, L. J. Action currents in stomach and intestine. Am. J. Physiol., 1922, 58, 476-493.

37. Alvarez, W. C. & Mahoney, L. J. Peristaltic rush as depicted in the electrogastrogram. Am. J. Physiol., 1924, 69, 226-228.

38. Alvarez, W. C. & Zimmerman, A. Movements of the stomach. Am. J. Physiol., 1928, 84, 261-270.

39. Anand, B. K. Influence of the internal environment on the nervous regulation of alimentary behavior. In M. A. B. Brazier (Ed.), Brain and Behavior, Vol. 2. The Internal Environment and Alimentary Behavior. Washington, D.C.: American Institute of Biological Sciences, 1962.

40. Anand, B. K. & Dua, S. Effect of electrical stimulation of the limbic system ("visceral brain") on gastric secretion and motility. Indian J. Med. Res., 1956, 44, 125-130.

41. Anderson, J. J., Bolt, R. J., Ullman, B. M. & Bass, P. Differential response to various stimulants in the body and antrum of the canine stomach. _Am. J. Dig. Dis._, 1968, _13_, 147-156.

42. Anderson, J. J., Bolt, R. J., Ullman, B. M. & Bass, P. Effect of bile and fat on gastric motility under the influence of various stimulants. _Am. J. Dig. Dis._, 1968, _13_, 157-167.

43. Anderson, W. F. Motility of the fasting stomach in health and disease. _Lancet_, 1943, _244_, 40-42.

44. Anderson, W. F. The effect of "Banthine" on balloon human gastric motility and on pulse rate. _J. Physiol._, London, 1953, _121_, 53.

45. Anderson, W. F. Action of "antrenyl," new antispasmodic agent, on gastric motility in man. _Glasgow Med. J._, 1954, _35_, 105-108.

46. Andersson, S. & Grossman, M. I. Profile of pH, pressure and potential difference at gastroduodenal junction in man. _Gastroenterology_, 1965, _49_, 364-371.

47. Androsov, N. S., Firsova, P. P. & Tolstykh, P. I. Significance of the electrogastrogram in evaluating gastric motor function in cancer. _Vopr. Onkol._, 1973, _19_, 39-43.

Comprehensive analysis of 100 electrogastrograms in patients with carcinoma of the stomach demonstrated reduced EGG biopotentials reflecting the tonus and peristaltic activity of the organ (mean, maximal and minimal biopotentials). A typical finding was a derangement of the rhythm of the wave pattern of the electrogastrographic biopotentials in various functional conditions of the stomach. Statistical processing of basic EGG indices enabled a quantitative evaluation of gastric biopotentials in relation to tumor localization.

48. Annegers, J. H. & Ivy, A. C. Method for determining effect of various agents on gastric evacuation in man. _Gastroenterology_, 1947, _8_, 711-716.

49. Annegers, J. H. & Ivy, A. C. The effect of dietary fat upon gastric evacuation in normal subjects. _Am. J. Physiol._, 1947, _150_, 461-465.

50. Antia, F. & Ivy, A. C. Effect of vagotomy on gastric secretion and emptying time in the dog. _Fed. Proc._, 1949, _8_, 5.

51. Antia, F., Rosiere, C. E., Robertson, C. & Grossman, M. I. Effect of vagotomy on gastric secretion and emptying time in dogs. Am. J. Physiol., 1951, 166, 470-479.

52. Apperly, F. L. Mixed effects of olive oil in clinical doses on gastric function. Gastroenterology, 1943, 1, 1127-1132.

53. Arguin, S. Modification of adrenalin effect with functional status of stomach. Proc. Soc. Exp. Biol. Med., 1927, 25, 97-98.

54. Arhan, P. & Faverdin, C. Method of functional exploration of digestive motility in pediatric surgery. Ann. Chir. Infant, 1971, 12, 197-206.

55. Arimori, M., Code, C. F., Schlegel, J. F. & Sturn, R. E. Electrical activity of the canine esophagus and gastroesophageal sphincter. Am. J. Dig. Dis., 1970, 15, 191.

56. Arkhipov, G. N. A method for recording gastric motility in rats with simultaneous collections of gastric juice. Patol. Fiziol. Eksp. Ter., 1965, 9, 79-81.

57. Armitage, A. K. & Dean, A. C. B. Function of the pylorus and pyloric antrum in gastric emptying. Gut, 1963, 4, 174-178.

58. Armitage, A. K. & Dean, A. C. B. The effects of pressure and pharmacologically active substances on gastric peristalsis in a transmurally stimulated rat stomach-duodenum preparation. J. Physiol., London, 1966, 182, 42-56.

59. Armstrong, W. McD. Sodium, potassium, and chloride in frog stomach muscle. Am. J. Physiol., 1964, 206, 469-475.

60. Asai, T. Developmental processes of reticulorumen motility in calves. Jap. J. Vet. Sci., 1973, 35, 239-252.

61. Atanassova, E. On the mechanism of correlation between the spike activities of the stomach and duodenum. Bull. Inst. Physiol. Bulg. Acad. Sci., 1970, 13, 229-242.

62. Atanassova, E., Jurukova, Z. & Zacheva, I. Regeneration of the smooth-muscle cells, a structural basis for restoration of the slow-potential rhythm in the distal part of the stomach after transection. J. Physiol., Paris, 1974, 68, 291-304.

63. Atkinson, M. Mechanisms protecting the gastroesophageal reflux: A review. Gut, 1962, 3, 1-15.

64. Atkinson, M., Edwards, D. A. W., Honour, A. J. & Rowlands, E. N. Comparison of cardiac and pyloric sphincters. Lancet, 1957, 2, 918-922.

65. Auer, J. The effect of severing the vagi or the splanchnics or both upon gastric motility in rabbits. Am. J. Physiol., 1910, 25, 334-344.

66. Aune, S. Intragastric pressure after vagotomy in man. Scand. J. Gastroenterol., 1969, 4, 447-452.

67. Awatsu, S. Regional differences in gastric motility. Jap. J. Smooth Muscle Res., 1974, 10, 155-157.

68. Axelsson, J. Dissociation of electrical and mechanical activity of smooth muscle. J. Physiol., London, 1961, 158, 381-398.

69. Aylett, P. Gastric emptying and change of blood glucose level, as affected by glucagon and insulin. Clin. Sci., 1962, 22, 171-178.

70. Aylett, P. The effects of changes in blood glucose levels on the gastric emptying rate. Am. J. Dig. Dis., 1968, 13, 1035-1046.

71. Babkin, B. P. Cerebral cortex and gastric motility. Edinburgh Med. J., 1950, 27, 419.

72. Babkin, B. P. The cerebral cortex and gastric motility. Gastroenterology, 1950, 14, 479-484.

73. Babkin, B. P. & Bornstein, M. B. The effect of swinging and of binaural galvanic stimulation on the motility of the stomach in dogs. Rev. Canad. Biol., 1943, 2, 336-349.

74. Babkin, B. P. & Friedman, M. H. T. Digestive system. Ann. Rev. Physiol., 1945, 7, 305-330.

75. Babkin, B. P. & Kite, W. C., Jr. Central and reflex regulation of motility of pyloric antrum. J. Neurophysiol., 1950, 13, 321-334.

76. Babkin, B. P. & Kite, W. C., Jr. Gastric motor effects of acute removal of ungulate gyrus and section of brain stem. J. Neurophysiol., 1950, 13, 335-342.

77. Babkin, B. P. & Speakman, T. J. Cortical inhibition of gastric motility. J. Neurophysiol., 1950, 13, 55-63.

78. Bachrach, W. H., Brody, M. & Drury, D. R. Multiple simultaneous recordings of gastrointestinal functions. Gastroenterology, 1951, 18, 609-612.

79. Bachrach, W. H., Mason, J. & Pollard, H. M. Fasting gastric motility in duodenal ulcer patients. Gastroenterology, 1952, 20, 234-237.

80. Bacon, C. The influence of psychologic factors upon gastro-intestinal distrubances: A symposium. Typical personality trends and conflicts in case of gastric distrubance. Psychoanal. Quart., 1934, 3, 558-573.

81. Bagby, R. & Fisher, B. Graded contractions in muscle strips and single cells from Bufo marinus (toad) stomach. Am. J. Physiol., 1973, 105-109.

82. Bailey, D. M. Inhibitory and excitatory effects of sympathomimetric amines on muscle strips from stomach of guinea pig. Br. J. Pharmacol., 1971, 41, 227-238.

83. Bailey, P. & Sweet, W. H. Effects on respiration, blood pressure and gastric motility of stimulation of orbital surface of frontal lobe. J. Neurophysiol., 1940, 3, 276-281.

84. Barado, S. F., Garcia, H. E., Lopez-Calderon, B. & Arias, V. E. Polygraphy in the diagnosis of rigid stomach. Rev. Esp. Enferm. Apar. Dig., 1970, 30, 266-280.

85. Barashkova, G. M. Gastrin, cholecystokinin-pancreozymin, secretin and the interaction of the motor function of the stomach, duodenum and bile-secreting apparatus. Fiziol. Zh. S.S.S.R., 1975, 61, 763-773.

86. Barboriak, J. J. & Meade, R. C. Effect of alcohol on gastric emptying in man. Am. J. Clin. Nutr., 1970, 23, 1151-1153.

87. Barclay, A. E. The Digestive Tract. Cambridge: Cambridge University Press, 1933.

88. Barker, G. R., Cochrane, G. M., Corbett, G. A., Hunt, J. N. & Roberts, S. K. Actions of glucose and potassium chloride on somoreceptors slowing gastric emptying. J. Physiol., London, 1974, 237, 183-186.

89. Baron, L. E., Curtis, G. M. & Haverfield, W. T. The effect of splanchnic resection on gastric motility in man. Am. J. Physiol., 1935, 113, 8.

90. Barowsky, H. Gastroscopic observations ot anticholinergic effect of pro-banethine on gastric motility and pyloric function. Am. J. Gastroenterol., 1955, 23, 557-563.

91. Barowsky, H., Schwartz, A. & Lister, J. New approach to management of peptic ulcer by inducing sustained gastric inactivity. Am. J. Gastroenterol., 1955, 24, 533-542.

92. Barr, L. Propagation in vertebrate visceral smooth muscle. J. Theoret. Biol., 1963, 4, 73-85.

93. Barr, L. & Dewey, M. M. Electrotonus and electrical transmission in smooth muscle. In C. F. Code (Ed.), Handbook of Physiology. Sect. 6, Alimentary Canal. Vol. IV. Motility. Washington, D.C.: American Physiological Society, 1968.

94. Barricks, M. E., Kohatsu, S. & Nelsen, T. S. Signal analysis of the telemetered electrogastrogram. Proc. Ann. Conf. Eng. Biol., 1968, 10, 38.

95. Bartlelmess, R. & Kunz, O. Comparative gastrocamera and radiographic studies in stomach candidiasis. Radiologe, 1973, 13, 388-393.

96. Bash, K. W. An investigation into a possible organic basis for the hunger drive. J. Comp. Psychol., 1939, 38, 109-134.

97. Basinska, H. The effect of resection of the vagus nerve on gastric motility and gastric glandular mucosa. Poznan. Tow. Przyj. Nauk. Wyolz. Lek., 1966, 34, 5-36.

98. Bass, P. Electrical activity of smooth muscle of the gastrointestinal tract. Gastroenterology, 1965, 49, 391-394.

99. Bass, P. The electrical and mechanical activity of the gastroduodenal junction. Jap. J. Smooth Muscle Res., 1970, 6, 93.

100. Bass, P. Relation of electrical activity to contraction. In L. Demling & R. Ottenjann (Eds.), Gastrointestinal Motility. New York: Academic Press, 1971.

101. Bass, P., Code, C. F. & Lambert, E. H. Electric activity at the gastroduodenal junction. Physiologist, 1959, 2, 8.

102. Bass, P., Code, C. F. & Lambert, E. H. Electrical activity of gastroduodenal junction. Am. J. Physiol., 1961, 201, 587-592.

103. Bass, P. & Weisbrodt, N. W. Current concepts in pharmacology of gastrointestinal motility. In L. Demling & R. Ottenjann (Eds.), Gastrointestinal Motility. New York: Academic Press, 1971.

104. Battelli, F. Action de diverses substances sur les mouvements de l'estomac, et innervation de cet organe. Compt. rend. Acad. d. sci., Paris, 1896, 122, 1568-1570.

105. Batygina, N. I. On the motor function of the gastric stump after resection of the stomach. Klin. Khir., 1966, 5, 63-65.

106. Baust, W. & Rohrwasser, S. Gastric motility and pH during natural human sleep. Pfluger's Arch. f. d. gesammt. Physiol., 1969, 305, 229-240.

107. Baziian, R. A. The effect of ligation of the major veins of the stomach on its motor and secretory functions under experimental conditions. Eksp. Khir. Anesteziol., 1966, 11 36-38.

108. Beani, L., Blanchi, C. & Crema, A. Vagal non-adrenergic inhibition of guinea-pig stomach. J. Physiol., 1970, 217, 259.

109. Beaumont, W. Experiments and Observations on the Gastric Juice and the Physiology of Digestion. Plattsburgh: F. P. Allen, 1833.

110. Beck, C. S. & Mason, G. R. Gastric peristalsis. A study of regional rates of contraction. Am. J. Surg., 1970, 119, 217-219.

111. Beck, C. S. & Osa, T. Membrane activity of the guinea pig gastric sling muscle--a nerve dependent phenomenon. Am. J. Physiol., 1971, 220, 1397-1403.

112. Beck, I. T., Rona, S. & Callegarini, U. The effect of gastric emptying on the (D) (plus)-xylose blood level time curve. Am. J. Dig. Dis., 1962, 7, 928-935.

113. Bedi, B. S. & Code, C. F. Pathway of coordination of postprandial, antral and duodenal action potentials. Am. J. Physiol., 1972, 222, 1295.

114. Bedi, B. S. & Kelly, K. A. Propagation of the canine gastric pacesetter potential. Gastroenterology, 1971, 60, 640.

115. Bedi, B. S., Kelly, K. A. & Holley, K. E. Pathways of propagation of the canine gastric pacesetter potential. Gastroenterology, 1972, 63, 288.

116. Bell, F. R. & Grivel, M. L. Proceedings: The role of the duodenal infusate on gastric emptying measured by the e.m.g. of gastric muscle. J. Physiol., London, 1974, 239, 48-49.

117. Bella, D. D. & Rognoni, F. Neurovegetative control of gastric motility in the isolated nerve-stomach preparation in the rat. J. Pharmacol. Exp. Therap., 1961, 134, 184-189.

118. Belousov, A. S., Levin, G. L. & Abramova, V. G. The effect of atropine and morphine on the evacuation of stomach content in patients with duodenal ulcer. Sov. Med., 1968, 31, 67-71.

119. Belousov, Iu. V. Proteolytic function of the stomach in chronic gastritis, peptic ulcer and functional secretory--motor diseases of the stomach in children. Pediatr. Akush. Ginekol., 1973, 5, 6-8.

120. Benmair, Y., Dreyfuss, F. & Fischel, B. Report on measurements of gastric function with a ferromagnetic tracer material. Eurocon Digest, (E-4-5), Lausanne, Switzerland, 1971.

121. Bennett, A. Effect of gastrin on isolated smooth muscle preparations. Nature, 1965, 208, 179-173.

122. Bennett, A. Relationship between in-vitro studies of gastrointestinal muscle and motility of the alimentary tract in vivo. Am. J. Dig. Dis., 1968, 13, 410-414.

123. Bennett, A., Bucknell, A. & Dean, A. C. B. The release of 5-hydroxytryptamine from the rat stomach in vivo. J. Physiol., London, 1966, 182, 57-65.

124. Bennett, A., Friedmann, C. A. & Vane, J. R. Release of prostaglandin E-1 from the rat stomach. Nature, 1967, 216, 873-876.

125. Bennett, A., Misiewicz, J. J. & Waller, S. L. Analysis of the motor effects of gastrin and pantagastrin on the human alimentary tract in vivo. Gut, 1967, 8, 470-474.

126. Bennett, A. & Whitney, B. A pharmacological study of the motility of the human gastrointestinal tract. Gut, 1966, 7, 307-316.

127. Bennett, D. R. & Bass, P. The action of inhalation anesthetics on the gastrointestinal tract. Handbook of Experimental Pharmacology, Modern Inhalation Anesthetics. Berlin: Springer, 1970.

128. Bennett, T. L. & Venables, J. F. The effect of emotions on the gastric secretion and motility in the human being. Br. Med. J., 1920, 2, 662.

129. Benoit, D., Fays, J., Hoeffel, J. C. & Gaucher, P. Gastric stasis after vagotomy. Acta Otorhinolaryngol. Belg., 1969, 23, 307-314.

130. Benoit, D., Fays, J., Hoeffel, J. C. & Gaucher, P. Gastric stasis after vagotomy. J. Chir., Paris, 1971, 101, 307-314.

131. Benzi, G. M., Arrigoni, E., Panceri, P., Parzarasa, R. & Berte, F. Action of xylamide on gastrointestinal motor activity in animals and animal organs. Digestion, 1973, 9, 150-165.

132. Bercovitz, Z. & Rogers, F. J. Contributions to the physiology of the stomach LV. The influence of the vagi on gastric tonus and motility in the turtle. Am. J. Physiol., 1921, 55, 323-338.

133. Berger, T. Studies on the gastric emptying mechanism in healthy persons and patients after partial gastrectomy. Acta Chir. Scand., 1969, 404, 1.

134. Berger, T., Kewenter, J. & Kock, N. G. Response to gastrointestinal pacing: antral, duodenal and jejunal motility in control postoperative patients. Ann. Surg., 1966, 164, 139-144.

135. Berger, W. & Barr, L. Evidence for electrical transmission of action potentials in vertebrate visceral smooth muscle. Physiologist, 1965, 8, 111.

136. Bernier, J. J., Bognel, J. C., Bognel, C. & Rambaud, J. C. Experimental study of the digestive toxicity of methotrexate. I. Effect on digestive motility. Rev. Fr. Etud. Clin. Biol., 1967, 12, 576-580.

137. Bertaccini, G. Action of caerulein on the motility of the biliary system and the gastrointestinal tract in man. Med. Chir. Dig., 1973, 2, 133-138.

138. Bertaccini, G. & Impicciatore, M. Action of bombesin on the motility of the stomach. Naunyn-Schmiedebergs Arch. Pharmakol. Exp. Patholog., 1975, 289, 149-156.

139. Bertrand, J. & Thillier, J. L. Action of metopimazine on gastric motility. Double-blind study using external digestive electromyography. Nouv. Presse Med., 1975, 4, 2319-2320.

140. Besancon, F. The current role of the graphic method in gastroenterology. Acta Gastroenterol. Belg., 1967, 30, 704-707.

141. Besancon, F., Baujat, J. P. & Debray, C. Gastroesphageal reflex and pH-graphic and electromanographic study. Semaine Hop. Paris, Paris, 1962, 38, 1569-1576.

142. Bevan, J. A. & Hughes, T. Inhibition of gastric contraction following stimulation of intrathoracic sensory endings. Arch. Int. Pharmacodyn., 1966, 161, 334-342.

143. Bisgard, J.D. & Nye, E. The influence of hot and cold application upon gastric and intestinal motor activity. Surgical Gynecology and Obstetrics, 1940, 71, 172.

144. Bilich, G. L., Donchenko, V. S. & Shedlovskaia, M. V. The effect of certain types of anesthesia on secretory-motor function of the stomach in children. Pediatriia, 1968, 47, 75-76.

145. Bingham, J. R., Ingelfinger, F. J. & Smithwick, R. H. Effects of sympathectomy on the motility of the human gastrointestinal and biliary tracts. Gastroenterology, 1950, 15, 6-13.

146. Bingham, J. R., Ingelfinger, F. J. & Smithwick, R. H. The effects of sympathectomy on abdominal pain in man. Gastroenterology, 1950, 15, 18-31.

147. Birchall, E. F., Fenton, P. F. & Pierce, H. H. Gastric emptying and intestinal absorption of dextrose solutions. Am. J. Physiol., 1946, 146, 610-612.

148. Birnbaum, D., Ben-Menachem, J. & Schwartz, A. The influence of oral diazepam on gastrointestinal motility. A preliminary report. Am. J. Proctol., 1970, 21, 263-266.

149. Bischop, C. R., Janzen, R. E., Landals, D. C., Manns, B. D., McCartney, D. J., Morgan, G. A., Pawlyshyn, V. P., Schienbein, A. J., Scigliano, B. W. & Taylor, K. L. Effects of rumen stasis on sulfamethazine blood levels after oral administration in sheep. Can. Vet. J., 1973, 14, 269-271.

150. Blair, E. L., Harper, A. A., Kidd, C. & Scratcherd, T. Post-activation potentiation of gastric and intestinal contractions in response to stimulation of the vagus nerve. J. Physiol., London, 1959, 148, 437-449.

151. Blair, E. L., Venables, C. W. & Wilkinson, J. Proceedings: Gastric relaxation in the dog in response to feedings. J. Physiol., London, 1974, 238, 72-73.

152. Blasi, A., Conticello, S., Mangiafico, F. & Marletta, F. Effects of olfactory stimulation on gastrointestinal motility. Clin. Otorinolaringoiatr., 1970, 22, 187-203.

153. Bloom, P. B., Filion, R. D., Stunkard, A. J., Fox, S. & Stellar, E. Gastric and duodenal motility, food intake and hunger measured in man during a 24-hr period. Am. J. Dig. Dis., 1970, 15, 710-723.

154. Boccadoro, B., Leonardi, L. & Calvari, A. R. Influence of sulpiride on gastrointestinal motility: Experimental radiological study in the dog. Arzneim. Forsch., 1975, 25, 905-916.

155. Boev, K. & Papazova, M. The influence of temperature on the slow potential from the smooth muscle of the stomach. Izv. Inst. Fiziol., (Sofiia), 1969, 12, 115-126.

156. Bogach, P. G. Motor function of the digestive tract and the process of digestion. Fiziol. Zh. S.S.S.R., 1967, 13, 635-644.

157. Bogach, P. G. & Groisman, S. D. Role of the extramural nervous system in the spreading of peristaltic waves in the stomach. Dikl. Akad. Nauk. S.S.S.R., 1966, 168, 708-711.

158. Bogach, P. G. & Groisman, S. D. The effect of transverse section of the stomach on the motor action of the fundal and pyloric regions. Eksp. Khir. Anesteziol., 1966, 11, 35.

159. Bogach, P. G., Groisman, S. D. & Chan-Xuy-Nga. Effect of different degrees of parasympathetic denervation of the stomach, duodenum and jejunum on their motor reactions evoked by insulin hypoglycemia. Fiziol. Zh. S.S.S.R., 1974, 60, 1446-1453.

160. Bogach, P. G., Kaplunenko, N. A., Chaichenko, G. M. & Milenov, K. Electric activity of the stomach and small intestine smooth muscles. Fiziol. Zh. S.S.S.R., 1971, 57, 276-283.

161. Bogach, P. G. & Koval, L. A. On the participation of the limbic cortex and the amygdaloid bodies in the regulation of the motor function of the gastrointestinal tract. Biull. Eksp. Biol. Med., 1968, 65, 11-15.

162. Bogach, P. G. & Koval, L. A. Motor activity of the stomach and intestines following hypothalamic administration of serotonin. Fiziol. Zh. S.S.S.R., 1975, 21, 37-40.

163. Bohr, D. E. Electrolytes and smooth muscle contraction. Pharmacol. Rev., 1964, 16, 85-111.

164. Boldireff, W. N. Die periodische Tatigkeit des Verdauungsapparates ausser der Verdauungszeit. Zentralbl. f. Physiol., 1904, 18, 489-493.

165. Boldireff, W. N. Le Travail periodique de l´appareil digestif en dehors de la digestion. Arch. d. Sci. Biol., 1905, 11, 1-5.

166. Boldireff, W. N. Bemerkungen zu den Artikeln von A. J. Carlson: Contributions to the physiology of the stomach. Zentralbl. f. Physiol., 1914, 28, 193-194.

167. Boldireff, W. N. Detailed account of the periodic activity of the digestive system. Ergeb. Physiol., 1929, 29, 485-645.

168. Boles, J. & Russell, R. W. Relations between the electrogastrogram and gastric ulceration during exposure to stress. Psychophysiology, 1970, 6, 404-410.

Relations between development of gastric ulceration and changes in motility of the stomach, as recorded by the electrogastrogram (EGG), were studied under conditions where subjects (rats) were exposed to physical (immobilization) and chemical (reserpine) stressors and to the two in combination. Following 36 hrs of the various treatments, all restraint + reserpine subjects were ulcerated; 60% of the restraint-only subjects had lesions; and only three of ten no-restraint subjects showed gastric damage. The same rank order was obtained for the mean number of lesions per group. There were no significant differences among groups in total acidity of the stomach at autopsy immediately after their experimental treatments. The restraint-only groups showed a trend toward increased gastric motility during the treatment period, while the restraint + reserpine groups showed a progressive decrease in motility, the divergence leading to statistically significant differences as exposure to the different stressors continued. The results do not support the hypothesis that the experimentally-produced lesions are related to an increase in gastric acicity; they do support the hypothesis that the ulceration is related to changes in gastric motility, more severe lesions appearing in subjects whose motility decreased during the treatment period.

169. Booligin, I. A. The relation between the cerebral cortex and the motor action of the digestive track. Akh. Biol. Nauk., 1934, 54, 65-78.

170. Booth, M., Hunt, J. N., Miles, J. & Murray, F. A. Comparison of gastric emptying in men and women. Lancet, 1957, 1, 657.

171. Borchers, E. Anteil des Nervus vagus an der motorischen Innervation des Magens im Hinblick auf die operative Therapie von Magenkrankheiten: Studien zur Physiologie und Pathologie der Magenbewegungen, sowie zu modernen Problemen der Magenchirurgie. Beitr. z. klin. Chir., 1921, 122, 547-622.

172. Borchers, E. Die Aussichten der Behandlung von Motilitatsstorungen des Magens durch Vagusunterbrechung. Deutsche Ztschr. f. Chir., 1921, 162, 19-28.

173. Borchers, E. Ueber die Resektion der kleinen Kurvatur und deren Folgen fur Magenform and Motilitat. Arch. f. klin. Chir., 1922, 122, 198-217.

174. Borison, H. L. & Wang, S. C. Physiology and pharmacology of vomiting. Pharmacol. Rev., 1953, 5, 193-230.

175. Bortoff, A. Digestion: Motility. Ann. Rev. Physiol., 1972, 34, 261.

176. Bortoff, A. & Davies, R. Myogenic transmission of antral slow waves across the gastroduodenal junction in situ. Am. J. Physiol., 1968, 215, 889.

177. Bortoff, A. & Weg, N. Transmission of electrical activity through the gastroduodenal junction. Am. J. Physiol., 1965, 208, 531-536.

178. Boullin, D. J. Observations on the significance of 5-hydroxytryptamine in relation to the peristaltic reflex of the rat. Br. J. Pharmacol., 1964, 23, 14-33.

179. Bourgill, J. E., Friedman, M. H. T., Patterson, T. L. & Sandweiss, D. J. Effect of intravenous injections of urine extracts on gastric motility. Gastroenterology, 1943, 1, 1049-1054.

180. Bowditch, H. P. Movements of the alimentary canal. Science, New Series, 1897, 5, 901-902.

181. Bowen, J. M. Effects of insulin hypoglycemia on gastrointestinal motility in the sheep. Am. J. Vet. Res., 1962, 23, 948-954.

182. Boyden, E. A. & Rigler, L. G. Initial emptying time of stomach in primigrovidal as related to evacuation of biliary tract. Proc. Soc. Exp. Biol. Med., 1944, 56, 200-201.

183. Bozler, E. The action potentials of visceral sheath muscle. Am. J. Physiol., 1938, 124, 502-510.

184. Bozler, E. Electrophysiological studies on the motility of the gastrointestinal tract. Am. J. Physiol., 1939, 127, 301-307.

185. Bozler, E. An analysis of the excitatory and inhibitory effects of sympathetic impulses and adrenaline on visceral smooth muscle. Am. J. Physiol., 1940, 130, 627-634.

186. Bozler, E. The action potentials accompanying conducted responses in visceral smooth muscles. Am. J. Physiol., 1942, 136, 553-560.

187. Bozler, E. The activity of the pacemaker previous to the discharge of a muscular impulse. Am. J. Physiol., 1942, 136, 543-552.

188. Bozler, E. The action potentials of the stomach. Am. J. Physiol., 1945, 144, 693-700.

189. Bozler, E. The action potentials of visceral smooth muscle. Fed. Proc., 1945, 4, 8.

190. Bozler, E. Conduction, automaticity, and tonus of visceral muscles. Experientia, 1948, 4, 213-218.

191. Bozler, E. Initiation of contraction in smooth muscle. Physiol. Rev., 1962, 42, 179.

192. Bozler, E., Calvin, M. E. & Watson, D. W. Exchange of electrolytes in smooth muscle. Am. J. Physiol., 1958, 195, 38-44.

193. Bozler, E. & Lavine, D. Permeability of smooth muscle. Am. J. Physiol., 1958, 195, 45-49.

194. Braasch, J. W. & Ellis, F. H., Jr. Gastroesophageal sphincter mechanism: Experimental study. Surgery, 1956, 39, 901-905.

195. Brady, J. V. Further comments on the gastrointestinal system and avoidance behavior. Psych. Rep., 1963, 12, 742.

196. Brady, J. V., Porter, R. W., Conrad, D. G. & Mason, J. Avoidance behavior and the development of gastroduodenal ulcers. J. Exper. Analytic. Beh., 1958, 1, 69-73.

197. Brandolini, G., Achilli, M. & Pacelli, L. Importance of the gastrointestinal motility in resuscitation. Acta Anaesthesiol (Padova), 1969, 20, 595-602.

198. Brandolini, G., Pacelli, L. & Liberta, E. Complications of post-operative gastrointestinal motility in gynecology after peridural anesthesia and general anesthesia. Acta Anaesthesiol (Padova), 1969, 20, 180-186.

199. Brink, B. M., Schlegel, J. F. & Code, C. F. The pressure profile of the gastroduodenal junctional zone in dogs. Gut, 1965, 6, 163-171.

200. Brody, D. A. & Quigley, J. P. Application of "inductograph" to registration of movements particularly of body structures such as pyloric sphincter. J. Clin. Med., 1944, 29, 863-867.

201. Brody, D. A. & Quigley, J. P. Intralumen pressures of stomach and duodenum in health and disease. Gastroenterology, 1947, 9, 570-575.

202. Brody, D. A. & Quigley, J. P. The production of waves of inhibition in the esophagealfundic region. Fed. Proc., 1948, 7, 13-14.

203. Brody, D. A. & Quigley, J. P. Registration of digestive tract intralumen pressures. Methods Med. Res., 1951, 4, 109.

204. Brody, D. A. & Quigley, J. P. Intralumen pressures of the stomach and duodenum in health and disease. Gastroenterology, 1968, 54, 768-769.

205. Brody, D. A., Werle, J. M., Meschan, I. & Quigley, J. P. Interlumen pressures of the digestive tract especially the pyloric portion. Am. J. Physiol., 1940, 130, 791-801.

206. Bromster, D., Carlberger, G. & Lundh, G. Measurement of gastric emptying-rate. Lancet, 1966, 2, 224-225.

207. Bromster, D., Carlberger, G. & Lundh, G. Measurement of gastric emptying rate using 131-IHSA. A methodological study in man. Scand. J. Gastroenterol., 1968, 3, 641-653.

208. Brooks, F. P. & Carr, D. H. Proceedings: Gastric acid and motor responses to electrical stimulation of afferent and efferent fibres in the cat's vagus. J. Physiol., London, 1975, 250, 19-18.

209. Brown, B. H., Ng, K. K., Kwong, N. K., Duthie, H. L., Whittaker, G. E. & Franko, C. I. Computer analysis and simulation of human gastroduodenal electrical activity. Med. Biol. Eng., 1971, 9, 305.

210. Brown, B. H., Smallwood, R., Duthie, H. L. & Stoddard, C. J. Intestinal smooth muscle electrical potentials recorded from surface electrodes. Med. Biol. Eng., 1975, 13, 97-103.

The source of electrical signals recorded from the human small intestine using surface electrodes was studied. Electrodes were placed on the trunks of 16 normal subjects, 3 pairs over the gastroduodenal area and a 4th pair placed across the chest and used to record respiration. Recordings were made after fasting and immediately following a meal. Fourier analysis and correlation methods were used to analyze the data. A significant frequency component at approximately 3 cycles/min (average 3.02 min) was found in 88% of the stretches of record analyzed. It is unlikely that the signals were an electrode artifact or that they were the result of mechanical artifacts from the gut. Evidence indicates that the signal arose from the basic eletrical rhythm of the stomach. In 28% of the recordings, consistent activity was found at a frequency of 10-12 cycles/min (average 10.8 cycles/min); this activity was found almost exclusively in the records made in the fasting state. Although this activity is near respiratory frequencies, simultaneous monitoring of respiration enabled the waveforms to be separated. The source of these rhythms is probably either the small or large bowel.

211. Brown, G. L. & Barry, R. C. Reversal of the gastric vagus. J. Physiol., 1932, 75, 213-225.

212. Brown, G. L. & McSwiney, B. A. The sympathetic innervation of the stomach. IV. Reversal of sympathetic action by luminal. J. Physiol., London, 1932, 74, 179-194.

213. Brown, G. L., McSwiney, B. A. & Wadge, W. J. Sympathetic innervation of the stomach. I. Effect of the stomach of the thoracic sympathetic trunk. J. Physiol., London, 1930, 70, 253-260.

214. Brown, J. C. Presence of a gastric motor stimulating property in duodenal extracts. Gastroenterology, 1967, 52, 225-229.

215. Brown, J. C., Cook, M. A. & Dryburgh, J. R. Molitin, a gastric motor activity-stimulating polypeptide: Final purification, amino acid composition, and c-terminal residues. Gastroenterology, 1972, 62, 401-404.

216. Brown, J. C., Johnson, L. P. & Magee, D. F. Effect of duodenal alkalinization on gastric motility. Gastroenterology, 1966, 50, 333-339.

217. Brown, J. C., Mutt, V. & Dryburgh, J. R. The further purification of motilin, a gastric motor activity stimulating polypeptide from the mucosa of the small intestine of hogs. Can. J. Physiol. Pharmacol., 1971, 49, 399.

218. Brown, J. C. & Parks, C. O. Effect on fundic pouch motor activity of stimulatory and inhibitory fraction separated from pancreaozymn. Gastroenterology, 1967, 53, 731-736.

219. Brownlee, M. & Kroopf, S. S. Letter: Metoclopramide for gastroparesis diabeticorum. New Engl. J. Med., 1974, 291, 1257-1258.

220. Brunemeier, S. H. & Carlson, A. J. Contributions to the physiology of the stomach. XIX. Reflexes from the intestinal mucosa to the stomach. Am. J. Physiol., 1915, 36, 191-195.

221. Brunemeter, E. H. & Carlson, A. J. Contributions to the physiology of the stomach. XIX. Reflexes from the intestinal mucosa to the stomach. Am. J. Physiol., 1915, 36, 191-195.

222. Brunner, H. The use of CR51 C13 in mixed test meals for simultaneous measurement gastric emptying, pancreatic enzyme secretion and bile acid secretion. Rev. Roum. Med. Intern., 1974, 11, 132-134.

223. Brunner, H., Northfield, T. C., Hofmann, A. F., Go, V. L. & Summerskill, W. H. Gastric emptying and secretion of bile acids, cholesterol and pancreatic anzymes during digestion. Duodenal perfusion studies in healthy subjects. Mayo Clin. Proc., 1974, 49, 851-860.

224. Budzinskii, L. A. & Samokhualov, V. I. Clinico-radiographic parallels in the evaluation of late results of treatment of a severe dumping syndrome with reconstructive jejunogastroplasty. Vestn. Khir., 1973, 111, 3-7.

225. Bulatao, E. & Carlson, A. J. Contributions to the physiology of the stomach. Influence of experimental changes in blood sugar level on gastric hunger contractions. Am. J. Physiol., 1924, 69, 107-115.

226. Bulbring, E. Correlation between membrane potential, spike discharge and tension in smooth muscle. J. Physiol., London, 1955, 128, 200-221.

227. Bulbring, E. The actions of humoral transmitters on smooth muscle. Br. Med. Bull., 1957, 13, 172-175.

228. Bulbring, E. Smooth muscle of the alimentary tract. In F. Avery Jones (Ed.), Modern Trends in Gastro-enterology, New York: Hoeber, 1958.

229. Bulbring, E. & Burnstock, G. Membrane potential changes associated with tachphylaxis and potentiation of the response to stimulating drugs in smooth muscle. Br. J. Pharmacol., 1960, 15, 611-624.

230. Bulbring, E. & Crema, A. Observations concerning the action of 5-hydroxytryptamine on the peristaltic reflex. Br. J. Pharmacol., 1958, 13, 444-457.

231. Bulbring, E., Crema, A. & Sapby, O. B. Observation concerning the action of 5-hydroxytryptamine on the peristaltic reflex. Br. J. Pharmacol., 1958, 13, 440-443.

232. Bulbring, E. & Gershon, M. D. Five-hydroxytryptamine participation in vagal relaxation of the stomach. J. Physiol., London, 1966, 186, 95-96.

233. Bulbring, E. & Gershon, M. D. Five-hydroxytryptamine participation in the vagal inhibitory innervation of the stomach. J. Physiol., 1967, 192, 823-846.

234. Bulbring, E., Kuriyama, H. & Twarog, B. Influence of sodium and calcium on spontaneous spike generation in smooth muscle. J. Physiol., London, 1962, 161, 48.

235. Bulbring, E. & Lin, R. C. Y. The effect of intraluminal application of 5-hydroxytryptamine and 5-hydroxytryptophan on peristalsis; the local production of 5-HT and its release in relation to intraluminal pressure and propulsive activity. J. Physiol., London, 1958, 140, 381-407.

236. Bulbring, E., Lin, R. C. Y. & Schofield, G. C. An investigation of the peristaltic reflex in relation to anatomical observations. Quart. J. Exp. Physiol., 1958, 43, 26-37.

237. Burns, T. D. & Reinke, D. A. Gastric motility inhibition and excitation elicited by vagal stimulation. Am. J. Physiol., 1971, 221, 226-229.

238. Burnstock, G., Campbell, G., Stachell, D. & Smythe, A. Evidence that adenosine triphosphate or a related nucleotide is the transmitter substance released by noradrenergic inhibitory nerves in the gut. Br. J. Pharmacol., 1970, 40, 668-688.

239. Burnstock, G., Holman, M. E. & Prosser, C. L. Electrophysiology of smooth muscle. Physiol. Rev., 1963, 43, 482-527.

240. Burnstock, G. & Holman, M. E. Effect of drugs on smooth muscle. Ann. Rev. Pharmacol., 1966, 6, 129-156.

241. Burnstock, G. & Merrillees, N. C. R. Structural and experimental studies on autonomic nerve endings in smooth muscle. In: Pharmacology of Smooth Muscle, 2nd Intern. Pharmacol. Meeting (Prague). Oxford: Pergamon, 1964, 1-17.

242. Burnstock, G. & Prosser, C. L. Conduction in smooth muscle. Comparative electrical properties. Am. J. Physiol., 1960, 199, 553-559.

243. Burnstock, G. & Prosser, C. L. Delayed repolarization in smooth muscles. Proc. Soc. Exp. Biol. Med., 1960, 103, 269-270.

244. Burnstock, G. & Prosser, C. L. Responses of smooth muscles to quick stretch; relation of stretch to conduction. Am. J. Physiol., 1960, 198, 921-925.

245. Burnstock, G. & Straub, R. W. A method for studying the effects of ions and drugs on the resting and action potentials in smooth muscle with external electrodes. J. Physiol., London, 1958, 140, 156-167.

246. Bury, K. D. & Jambunathan, G. Effects of elemental diets on gastric emptying and gastric secretion in man. Am. J. Surg., 1974, 127, 59-64.

247. Bustos, F. L., Gonzalez, E. & Marzi, A. Relationship between intraduodenal lipolysis and gastric emptying. Acta Gastroenterol. Latinoam, 1973, 5, 57-60.

248. Butvin, G. K. State of emptying-motor function of the stomach and intestine following jejunogastroplasty. Brach Delo, 1972, 8, 80-84.

249. Bveno, L. & Ruckenbusch, Y. The cyclic motility of the omasum and its control in sheep. J. Physiol., 1974, 238, 295-312.

250. Calberger, G., Barany, F., Lundh, G. & Moberg, S. Emptying of the stomach and rate of passage in the upper part of the small intestine. Nord. Med., 1971, 86, 884.

251. Calderon, M., Sonnemaker, R. E., Hersh, T. & Burdine, J. A. 99M TC-Human albumin microspheres (HAM) for measuring the rate of gastric emptying. Radiology, 1971, 101, 371-374.

252. Califano, G. & Zarrilli, L. Gastro-esophageal junction. Mechanisms of closure. Chir. Patol. Sper., 1972, 20, 243-261.

253. Campbell, G. The inhibitory nerve fibres in the vagal supply to the guinea-pig stomach. J. Physiol., London, 1966, 185, 600-612.

254. Campbell, G. The autonomic innervation of the stomach of a toad (Bufo marines). Comp. Biochem. Physiol., 1969, 31, 693-706.

255. Campbell, K. N. & Patterson, T. L. The influence of upper urinary tract distention on gastric hunger motility in the dog. Am. J. Physiol., 1939, 126, 456-457.

256. Cannon, W. B. The movements of the stomach studied by means of the roentgen rays. Am. J. Physiol., 1898, 1, 13-14.

257. Cannon, W. B. The movements of the stomach studied by means of the roentgen rays. Am. J. Physiol., 1898, 1, 359-382.

258. Cannon, W. B. Auscultation of the rhythmic sounds produced by the stomach and intestines. Am. J. Physiol., 1905, 14, 339.

259. Cannon, W. B. The motor activities of the stomach and small intestine after splanchnic and vagus section. Am. J. Physiol., 1906-7, 17, 429-442.

260. Cannon, W. B. The Mechanical Factors of Digestion. New York: Longmans, 1911.

261. Cannon, W. B. The motor activities of the stomach and small intestine after splanchnic and vagus section. Am. J. Physiol., 1906, 17, 429-442.

262. Cannon, W. B. The influence of emotional states on the functions of the alimentary canal. Am. J. Med. Sci., 1909, 137, 480-487.

263. Cannon, W. B. The importance of tonus for the movements of the alimentary canal. Arch. Int. Med., 1911, 8, 417-426.

264. Cannon, W. B. The nature of gastric peristalsis. Am. J. Physiol., 1911, 29, 250-266.

265. Cannon, W. B. Bodily Changes in Pain, Hunger, Fear and Rage. New York and London: D. Appleton and Company, 1915.

266. Cannon, W. B. Hunger and thirst. In C. Murchison (Ed.), Handbook of General Experimental Psychology. Worcester, Mass.: Clark University Press, 1934, 247-263.

267. Cannon, W. B. & Lieb, C. The receptive relaxation of the stomach. Am. J. Physiol., 1912, 29, 267-273.

268. Cannon, W. B. & Murphy, F. T. The movements of the stomach and intestine in some surgical conditions. Ann. Surg., 1906, 43, 512-536.

269. Cannon, W. B. & Washburn, A. L. An explanation of hunger. Am. J. Physiol., 1912, 29, 441-454.

270. Capper, W. M., Airth, G. R. & Kilby, J. O. A test for pyloric regurgitation. Lancet, 1966, 2, 621-623.

271. Carlson, A. J. Contributions to the physiology of the stomach. I. The character of the movements of the empty stomach in man. Am. J. Physiol., 1912, 31, 151-168.

272. Carlson, A. J. The character of the movements of the empty stomach in man. Am. J. Physiol., 1912, 31, 151-168.

273. Carlson, A. J. The relation between the contractions of the empty stomach and the sensation of hunger. Am. J. Physiol., 1912, 31, 175-192.

274. Carlson, A. J. Contributions to the physiology of the stomach. II. The relation between the contractions of the empty stomach and the sensation of hunger. Am. J. Physiol., 1913, 31, 175-192.

275. Carlson, A. J. Contributions to the physiology of the stomach. III. The contractions of the empty stomach inhibited reflexly from the mouth. Am. J. Physiol., 1913, 31, 212-222.

276. Carlson, A. J. Contributions to the physiology of the stomach. V. The influence of stimulation of the gastric mucosa on the contractions of the empty stomach (hunger contractions) in man. Am. J. Physiol., 1913, 32, 245-263.

277. Carlson, A. J. Contributions to the physiology of the stomach. VI. A study of the mechanisms of the hunger contractions of the empty stomach by experiments on dogs. Am. J. Physiol., 1913, 32, 369-388.

278. Carlson, A. J. Contributions to the physiology of the stomach. VII. The inhibitory reflexes from the gastric mucosa. Am. J. Physiol., 1913, 32, 389 397.

279. Carlson, A. J. Contributions to the physiology of the stomach. VIII. The hunger contractions of the empty stomach during prolonged starvation (man, dog). Am. J. Physiol., 1914, 33, 95-118.

280. Carlson, A. J. The Control of Hunger in Health and Disease. Chicago: University of Chicago Press, 1916.

281. Carlson, A. J., Boyd, T. E. & Pearcy, J. F. Studies on the visceral sensory nervous system. XIV. The reflex control of the cardia and lower esophagus in mammals. Arch. Int. Med., 1922, 30, 409-433.

282. Carlson, A. J. & Ginsburg, H. Contributions to the physiology of the stomach. XXX. The tonus and contractions of the empty stomach of infants with congential pyloric stenosis, pylorospasm and chronic vomiting (merycism). Am. J. Physiol., 1916, 39, 310-312.

283. Carlson, A. J. & Litt, S. Studies on the visceral nervous system on the reflex control of the pylorus. Arch. Int. Med., 1924, 33, 281-291.

284. Carlson, A. J., Smith, E. A. & Gibbons, I. The action of choline on the alimentary canal of intact dogs. Am. J. Physiol., 1927, 81, 431-435.

285. Carlson, G. M., Ruddon, R. W., Hug, C. C. & Bass, P. Effects of nicotine on gastric antral and duodenal contractile activity in the dog. J. Pharmacol. Exp. Therap., 1970, 172, 367.

286. Carlson, G. M., Ruddon, R. W., Hug, C. C., Schmiege, S. K. & Bass, P. Analysis of the site of nicotine action on gastric antral and duodenal contractile activity. J. Pharmacol. Exp. Therap., 1970, 172, 377-383.

287. Carlson, H. C., Code, C. F. & Nelson, R. A. Motor action of the canine gastroduodenal junction: A cineradiographic, pressure and electric study. Am. J. Dig. Dis., 1966, 11, 155-172.

288. Carter, D. C., Whitfield, H. N. & MacLeod, I. B. The effect of vagotomy on gastric adaptation. Gut, 1971, 13, 874-879.

289. Casapo, A. Molecular structure and function of smooth muscle. In G. H. Bourne (Ed.), The Structure and Function of Muscle, Vol. I. New York: Academic Press, 1960.

290. Cascao De Anciaes, J. H. Insulin and gastric function. Arch. f. Verdauungskr., 1928, 42, 377-382.

291. Cassano, C. C., Amoruso, M., Arullani, P. & Badalamenti, G. Research on total dna partial gastrointestinal transit times. Arch. Ital. Mal. Appar. Dig., 1967, 34, 290-293.

292. Castro-Rial, M., Salamanca, F. E., Jr., Porrua, J. M. L. & Sterling, P. T. Accion de la cortisona en el functionalismo gastrico en los Perros hipofisectomizados III communicacion. Arch. Med. Exptl., Madrid, 1954, 17, 503-530.

293. Cate, J. Contribution a la physiologie comparee du canal stomaco-intestinal II. Mouvements rhythmigues spontanes de l´estomac isole d´Helix pomatia. Arch. Neerl. Physiol., 1924, 9, 199-212.

294. Cathcart, J. P. S. The role of emotions in the production of gastro-intestinal disturbances. Can. Med. Assoc. J., 1946, 55, 465-470.

295. Cavell, B. Gastric emptying in infants. Acta Paediatr. Scand., 1971, 60, 370-371.

296. Cerqua, S. The part played by splanchnic innervation in emptying of the stomach. J. Physiol., London, 1935, 84, 232-249.

297. Chaddock, T. E., Carlson, G. M. & Hamilton, C. L. Gastric emptying of a nutritionally balanced liquid diet in the rhesus monkey. In E. E. Daniel et al. (Eds.), Proc. 4th International Symposium on Gastrointestinal Motility. Vancouver: Mitchel Press.

298. Chang, C. A., McKenna, R. D. & Beck, I. T. Gastric emptying rate of the water and fat phases of a mixed test meal in man. Gut, 1968, 9, 420-424.

299. Chang, H. C. The action of choline, adrenaline and ephedrine on gastric motility. Chin. J. Physiol., 1929, 3, 229-248.

300. Chapman, W. P. & Wilkins, E. W., Jr. Van Hueber, E. Electrical stimulation of the central end of the vagus nerve in the unanesthetized dog. Method and effects on gastro-intestinal, respiratory, and circulatory systems. Surg. Gynecol. Obstet., 1954, 98, 353-362.

301. Chapman, W. P., Wyman, S. M., Mora, L. O., Gillis, M. A. & Jones, C. M. Barium studies of comparative action of vantrine, tincture of belladonna and placebos on motility of gastrointestinal tract in man. Gastroenterology, 1953, 23, 234-243.

302. Chaudhuri, T. K. Use of 99MTC-DTPA for measuring gastric emptying time. J. Nucl. Med., 1974, 15, 391-395.

303. Chey, W. Y., Hitanant, S., Hendricks, J. & Lorber, S. H. Effect of secretin and cholecystokinin on gastric emptying and gastric secretion in man. Gastroenterology, 1970, 58, 820.

304. Chey, W. Y., Kosay, S., Hendricks, J., Brauerman, S. & Lorber, S. H. Effect of secretin on motor activity of stomach and Heidenhain pouch in dogs. Am. J. Physiol., 1969, 217, 848.

305. Child, G. P. Pharmacological and clinical studies with antispasmodics and local anesthetics on upper digestive tract. Gastroenterology, 1951, 19, 126-136.

306. Chinn, H. I. & Wang, S. C. Locus of emetic action following irradiation. Proc. Soc. Exp. Biol. Med., 1954, 85, 472-474.

307. Chiu, C. Y. & Lee, C. Y. The effect of some adrenergic blocking agents on the inhibition of peristalsis caused by sympathomimetic amines. J. Formosan Med. Assoc., 1961, 60, 1128.

308. Chocholac, J. Mucosal valvular closure of the pylorous. Cesk. Radiol., 1966, 20, 217-231.

309. Christensen, J. The adrenergie nerves and gastrointestinal smooth muscle function. Gastroenterology, 1968, 55, 135-138.

310. Christensen, J. Pharmacologic identification of the lower esophageal sphincter. J. Clin. Invest., 1970, 49, 681-691.

311. Christensen, J. The controls of gastrointestinal movements: Some old and new views. New Engl. J. Med., 1971, 285, 85-98.

312. Christensen, J. The physiology of gastrointestinal transit. Med. Clin. North Am., 1974, 58, 1165-1180.

313. Chvasta, T., Weisbrodt, N. W. & Cooke, A. R. The effect of pentagastrin and secretin on gastric emptying in the dog. Clin. Res., 1971, 19, 656.

314. Cintra do Prado, F., Ferraz, W. G., Lopes, E. R. & Metzner, B. S. Effect of meprobamate on gastric activity. Am. J. Dig. Dis., 1961, 6, 833-840.

315. Clark, C. G. & Vane, J. R. The cardiac sphincter in the cat. Gut, 1961, 2, 252-262.

316. Clarke, R. J. & Williams, J. A. The effect of antral denervation and pyloroplasty on gastric emptying and "dumping." Br. J. Surg., 1972, 59, 903.

317. Claure, H., Calderon, C., Braunschweig, T. & Diaz, G. Gastric emptying time in normal subjects using CR-51 and a gamma camera. Rev. Med. Chil., 1974, 102, 936-938.

318. Clemenacon, G. Antrum motility disorder. A gastrocinematographic study using super-8-film. Z. Gastroenterol., 1973, 11, 97-102.

319. Cobb, J. S., Bank, S., Marks, I. N. & Louw, J. H. Gastric emptying after vagotomy and pyloroplasty. Relation to some postoperative sequelae. Am. J. Dig. Dis., 1971, 16, 207-215.

320. Coddington, R. W. Study of an infant with a gastric fistula and her normal twin. Psychosom. Med., 1968, 30, 172.

321. Code, C. F. The digestive system. Ann. Rev. Physiol., 1953, 15, 106-138.

O 322. Code, C. F. Handbook of Physiology. Sect. 6, Alimentary Canal. Vol. IV. Motility. American Physiological Society, Washington, D. C.: 1968.

323. Code, C. F. & Carlson, H. C. Motor activity of the stomach. In C. F. Code (Ed.), Handbook of Physiology. Sect. 6, Alimentary Canal. Vol. IV. Motility. Washington, D.C.: American Physiological Society, 1968.

324. Code, C. F., Hightower, N. C. & Morlock, C. G. Motility of the alimentary canal in man; review of recent studies. Am. J. Med., 1952, 13, 328-351.

325. Code, C. F., Kelley, M. L., Jr., Olsen, A. M., Ellis, F. H., Jr. & Schlegel, J. F. Hypertensive gastroesophageal sphincter. Proc. Staff Meetings Mayo Clinic, 1960, 35, 391-399.

326. Code, C. F. & Marlett, J. A. Canine tachygastria. Mayo Clin. Proc., 1974, 49, 325-332.

327. Code, C. F. & Marlett, J. A. The interdigestive myo-electric complex of the stomach and small bowel of dogs. J. Physiol., London, 1975, 246, 289-309.

328. Code, C. F. & Schlegel, J. F. The pressure profile of the gastroesophageal sphincter in man: An improved method of detection. Proc. Staff Meetings Mayo Clinic, 1958, 33, 406-414.

329. Code, C. F., Szurszewski, J. H. & Kelly, K. A. A concept of motor control by the pacesetter potential in the stomach and small bowel. Am. J. Dig. Dis., 1971, 16, 601.

330. Code, C. F., Szurszewski, J. H., Kelly, K. A. & Smith, I. B. A concept of control of gastrointestinal motility. In C. F. Code (Ed.), Handbook of Physiology. Sect. 6, Alimentary Canal. Vol. IV. Motility. Washington, D.C.: American Physiological Society, 1968.

331. Cohen, S. Disorders of gastrointestinal motor and sphincter function. Trans. Stud. Coll. Physicians, 1975, 42, 359-392.

332. Cohen, S. & Harris, L. D. Lower esophageal sphincter pressure as an index of lower esophageal sphincter strength. Gastroenterology, 1970, 58, 157-162.

333. Colcher, H., Goodman, E. N. & Katz, G. M. Gastrointestinal mechanical and electrical activities after gastrectomy. Am. J. Gastroenterol., 1959, 31, 408.

334. Cole, L. G. The complexmotor phenomena of various types of unobstructed gastric peristalsis. Arch. Roentg. Ray., 1911, 16, 242-247.

335. Cole, L. G. A radiographic study of the pylorus and duodenum with and without artificial dilatation of the duodenum. Arch. Roentg. Ray., 1912, 16, 425-428.

336. Cole, L. G. Serial radiography in the differential diagnosis of carcinoma of the stomach, gall bladder infection, and gastric or duodenal ulcer. Arch. Roentg. Ray., 1912-1913, 17, 172-181.

337. Cole, L. G. Relation of lesions of the small intestine to disorders of the stomach and cap as observed roentgenologically. Am. J. Med. Sci., 1914, 148, 92-118.

338. Cole, L. G. Motor phenomena of the stomach, pylorus and cap observed roentgenographically. Am. J. Physiol., 1917, 42, 618-619.

339. Colmer, M. R., Owen, G. M. & Shields, R. Pattern of gastric emptying after vagotomy and pyloroplasty. Br. Med. J., 1973, 78, 669-675.

340. Comino, E., Fava, C., Ragni, G. & Stoppa, I. M. Effects of carvlein on gastrointestinal motility. Minerva Med., 1973, 64, 4686-4691.

341. Connell, A. M. Measurement of intraluminal pressures. Problems of methodology and interpretation, and analysis of records. Am. J. Dig. Dis., 1968, 13, 397-409.

342. Connell, A. M. Modern views on physiology. X. The movements of the alimentary tract. Practitioner, 1969, 203, 517-575.

343. Connell, A. M. Clinical aspects of motility. Med. Clin. North Am., 1974, 58, 1201-1216.

344. Connell, A. M. & Logan, C. J. H. The role of gastrin in gastroileocolic responses. Am. J. Dig. Dis., 1967, 12, 277.

345. Connell, A. M. & Rowlands, E. N. Clinical applications of radio pills for measuring gastrointestinal pressure. In Proc. 3rd Intern. Conf. Med. Electr., London (Part I). Springfield, Thomas, 1960.

346. Connell, A. M. & Rowlands, E. N. Wireless telemetering from the digestive tract. Gut, 1960, 1, 266.

347. Connell, A. M., Texter, E. C., Jr. & Vantrappen, G. Classification and interpretation of motility records. Am. J. Dig. Dis., 1965, 10, 481.

348. Conrad, R. A. Effect of gamma radiation on gastric empyting time in the dog. J. Appl. Physiol., 1956, 9, 234.

349. Contejean, C. Action des nerfs pneumogastrique et grand sympatheque sur l´estomac chez les batraciens. Arch. Physiol. Norm. Pathol., 1892, 4, 640-650.

350. Contejean, C. Action des nerfs pneumogastrique et grand sympathique sur l´estomac chez les batraciens. Arch. de physiol. norm. et path., Paris, 1892, 4, 640-650.

351. Contejean, C. Innervation de l´estomac chez les batraciens. Compt. rend. de la Soc. de Biol., 1896, 3, 1050-1051.

352. Cook, H. M. & Riley, J. L. Radio telemetering capsule for recording stomach motility in ruminants. Cornell Vet., 1970, 60, 317-329.

353. Cook, M. A., Kolodej, A. & Kowalewski, K. Electrical and mechanical activity of canine stomach after preservation. Surg. Gynecol. Obstet., 1974, 11, 225-230.

354. Cooke, A. R. Duodenal acidification: Role of first part of duodenum in gastric emptying and secretion in dogs. Gastroenterology, 1974, 67, 35-41.

355. Cooke, A. R. Control of gastric emptying and motility. Gastroenterology, 1975, 68, 804-816.

356. Cooke, A. R., Cavasta, T. E. & Weisbrodt, N. W. Effect of pentagastrin on emptying and electrical and motor activity of the dog stomach. Am. J. Physiol., 223, 934.

357. Cooke, A. R. & Christensen, J. Motor functions of the stomach. In M.M.H. Sleisenger and J.S. Fordtran (Eds.), Gastrointestinal Disease. Philadelphia: W.B. Saunders, 1973.

358. Cooke, A. R. & Moulang, J. Control of gastric emptying by amino acids. Gastroenterology, 1972, 62, 528-532.

359. Costell, D. O. & Harris, L. D. Hormonal control of gastroesophageal-sphincter strength. New Engl. J. Med., 1970, 282, 886-889.

360. Courtade, D. & Guyon, J. F. Contribution a l'etude de l innervation motrice de l'estomac. J. de physiol. et de path. gen., Paris, 1899, 1, 38-46.

361. Couturier, D., Roze, C. & Debray, C. Motor activity of the duodenum in man: Correlation with antral contraction and electrical activity. J. Physiol., Paris, 1970, 62, 387-405.

362. Couturier, D., Roze, C., Paolaggi, J. A. & Debray, C. Activite electrique de l'estomac humain normal; variations topologiques. Compt. Rend. Acad. Bulgare Sci., Paris, 1970, 270, 2817-2820.

363. Couturier, D., Roze, C., Paolaggi, J. A. & Debray, C. Measurement of the characteristic parameters of the normal gastric electromyogram. Variations according to electrode location. Biol. Gastroenterol., Paris, 1971, 1, 11-22.

364. Couturier, D., Roze, C., Paolaggi, J. A. & Debray, C. Electrical activity of the normal human stomach: A comparative study of recordings obtained from the serosal and mucosal sides. Am. J. Dig. Dis., 1972, 17, 969-976.

365. Couturier, D., Roze, C., Vasconcellos, D., Accary, J. P. & Debray, C. Effects of acid in stomach and duodenum upon the gastric myoelectrical activity in man. Digestion, 1973, 9, 502-513.

366. Cowgill, C. R., Deuel, H. J., Jr., Plummer, N. & Messer, F. C. Studies in the physiology of vitamins. IV. Vitamin B in relation to gastric motility. Am. J. Physiol., 1926, 77, 389-401.

367. Cowley, D. J., Vernon, P., Jones, T., Glass, H. I. & Cox, A. G. Gastric emptying of solid meals after truncal vagotomy and pyloroplasty in human subjects. Gut, 1972, 13, 176-181.

368. Cox, W. M. & Mueller, A. J. Intravenously administered crystalline amino acids in causation of vomiting in dogs. Fed. Proc., 1944, 3, 56.

369. Crabbe, P. A. & Heremans, J. F. The distribution of immunoglobulin-containing cells along the human gastrointestinal tract. Gastroenterology, 1966, 51, 305-316.

370. Craver, B. N., Barrett, W., Cameron, A., Earl, A. E. & Yonkman, F. F. Pharmacological properties of new antispasmodic N. N.-dimethlylthymyloxyacetamide hydrochloride. Am. J. Dig. Dis., 1951, 18, 241-245.

371. Creamer, B., Cottom, D. G. & Hindle, W. Pyloric stenosis: A physiological exploration. Lancet, 1961, 1, 1655-1657.

372. Crider, J. O. & Thomas, J. E. Study of gastric emptying with pylorus open. Am. J. Dig. Dis. Nutr., 1937, 4, 295-300.

373. Crider, J. O. & Thomas, J. E. A further study of the inhibitory effect on gastric peristalsis of the products of protein digestion. Am. J. Physiol., 1938, 123, 44-45.

374. Crohn, B. B. & Wilensky, A. O. Studies in the variations of the tonus of the gastric musculature in health and disease. Arch. Int. Med., 1917, 20, 145-160.

375. Crone, R. S. & Ardran, G. M. The effect of morphine sulfate on gastric motility: Some radiologic observations in man. Gastroenterology, 1957, 32, 88-95.

376. Cummins, A. J. Gastric analysis: Methods and interpretations. In C.M. Thompson, (Ed.), The Stomach. New York: Grune & Stratton, 1967.

377. Cummins, A. J., Schapiro, H. & Meeker, P. C. Effect of secretin on gastric motility. Am. Surg., 1966, 32, 187-189.

378. Cutter, M. & Hein, P. A technique for recording gastric activity in a freely moving animal. Psychophysiology, 1968, 4, 499.

379. Dacka, E. & Smajkiewicz, L. Effect of vagotomy with procedures facilitating the gastric emptying on the motility of bile ducts. Pol. Przegl. Chir., 1973, 45, 1427-1430.

380. Dacka, E. & Smajkiewicz, L. Effect of vagotomy on motor activity of the stomach and intestines. Pol. Przegl. Chir., 1974, 46, 131-135.

381. Dagradi, A. E., Stempiln, S. J., Seifer, H. W. & Weinberg, J. A. Terminal esophageal (vestibular) spasms after vagotomy. Arch. Surg., 1962, 85, 955-968.

382. Dahlgren, S. & Selking, O. Gastrointestinal motility in healthy persons--study with the endoradiosonde. Nord. Med., 1970, 84, 928-929.

383. Dahlgren, S. & Selking, O. Motility of the human digestive tract under resting conditions and after ingestion of food. A study with endoradiosondes. Ups. J. Med. Sci., 1972, 77, 167-174.

384. Dahm, K. & Sapounov, S. Tonus and motility of the stomach and duodenum following selective vagotomy and pyloroplasty. Dtsch. Med. Wochenschr., 1969, 94, 1513-1518.

385. Dale, H. H. & Feldberg, W. The chemical transmitter of vagus effects to the stomach. J. Physiol., London, 1934, 81, 320-334.

386. van Dam, A. P. Gastric emptying time studied by means of radionuclide scanning. Rev. Roum. Med. Intern., 1974, 11, 121-125.

387. van Dam, A. P. The gamma camera in clinical evaluation of gastric emptying. Radiology, 1974, 110, 155-157.

388. Danhof, I. E. Gastrointestinal motility in man: A study of propenzdate. Am. J. Gastroenterol., 1967, 48, 185-197.

389. Danhof, I. E. Pentazocine effects on gastrointestinal motor functions in man. Am. J. Gastroenterol., 1967, 48, 295-310.

390. Danhof, I. E., Blackmore, W. P. & Upton, G. L. Effect of pentazocine on gastric emptying and small intestinal propulsive motility in the rat. Toxicology and Applied Pharmacology, 1966, 9, 356-362.

391. Daniel, E. E. Effect of drugs on contractions of vertebrate smooth muscle. Ann. Rev. Pharmacol., 1964, 4, 189-222.

392. Daniel, E. E. Electrical and contractile responses of the pyloric region to adrenergic and cholinergic drugs. Can. J. Physiol. Pharmacol., 1965, 44, 951-979.

393. Daniel, E. E. The electrical and contractile activity of the pyloric region in dogs and the effects of drugs. Gastroenterology, 1965, 49, 403-418.

394. Daniel, E. E. Electrical and contractile responses of the pyloric region to adrenergic and cholinergic drugs. Can. J. Physiol. Pharmacol., 1966, 44, 951-979.

395. Daniel, E. E. Further studies on the pharmacology of the pyloric region, analysis of the effects of intra-arterial histamine, serotonin, phenyl diguanide, morphine, and other drugs on the antrum and duodenal bulb. Can. J. Physiol. Pharmacol., 1966, 44, 981-1019.

396. Daniel, E. E. Pharmacology of the gastrointestinal tract. In C. F. Code (Ed.), Handbook of Physiology. Sect. 6, Alimentary Canal. Vol. IV. Motility. Washington, D.C.: American Physiological Society, 1968.

397. Daniel, E. E. The electrical activity of the alimentary tract. Am. J. Dig. Dis., 1968, 13, 297-319.

398. Daniel, E. E. Digestion: Motor function. Ann. Rev. Physiol., 1969, 31, 203-226.

399. Daniel, E. E. Further studies on the pharmacology of the pyloric region: Analysis of the effects of intra-arterial histamine, serotonin, morphine and other drugs on the antrum. Can. J. Physiol. Pharmacol., in press.

400. Daniel, E. E. A conceptual analysis of the pharmacology of gastrointestinal motility. In P. Holton (Ed.), Internat. Encyclop. of Pharmacal. and Therap., Section 39a. Pharmacology of Gastrointestinal Motility and Secretion. Vol. II, Oxford: Pergamon Press, 1973.

401. Daniel, E. E. & Chapman, K. M. Electrical activity of the gastrointestinal tract as an indication of mechanical activity. Am. J. Dig. Dis., 1963, 8, 54-102.

402. Daniel, E. E. & Holman, M. E. Third international symposium on gastrointestinal motility. Stockholm, September, 1971. Summary comments on session on electrophysiology. Am. J. Dig. Dis., 1972, 17, 287-288.

403. Daniel, E. E. & Irwin, J. Electric activity of gastric musculature. In C. F. Code (Ed.), Handbook of Physiology. Sect. 6, Alimentary Canal. Vol. IV. Motility. Washington, D.C.: American Physiological Society, 1968.

404. Daniel, E. E. & Irwin, J. Electrical activity of the stomach and upper intestine. Am. J. Dig. Dis., 1971, 16, 602-610.

405. Daniel, E. E., Robinson, K., Duchon, G. & Henderson, R. M. The possible role of close contacts (nexuses) in the propagation of control electrical activity in the stomach and small intestine. Am. J. Dig. Dis., 1971, 16, 611-622.

406. Daniel, E. E. & Sarna, S. K. Organization of vagal fibers to control gastric motility. Gastroenterology, 1975, 68, 879.

407. Daniel, E. E., Sehdev, H. & Robinson, K. Mechanisms for activation of smooth muscle. Physiol. Rev., 1962, 5, 228-260.

408. Daniel, E. E. & Wiebe, G. E. Transmission of reflexes arising on both sides of the gastroduodenal junction. Am. J. Physiol., 1966, 211, 634-642.

409. Darnis, F. & Streichenberger, G. Quality of results obtained in digestive pathology by acting upon motor, secretory and hormonal phenomena at the same time. Therapeutique, 1973, 49, 721-724.

410. Davenport, H. W. Digestive system. Ann. Rev. Physiol., 1959, 21, 183-210.

411. Davenport, H. W. Stimulation of gastric motility by acid. Gastroenterology, 1967, 52, 198-204.

412. Davenport, H. W. Physiology of the Digestive Tract. Chicago: Year Book, 1971.

413. Davies, W. T., Griffith, G. H., Owen, G. M. & Shields, R. The effect of vagotomy and drainage operations on the rate of gastric emptying in duodenal ulcer patients. Br. J. Surg., 1974, 61, 509-515.

414. Davies, W. T., Kirkpatrick, J. R., Owen, G. M. & Shields, R. Gastric emptying in atrophic gastritis and carcinoma of the stomach. Scand. J. Gastroenterol., 1971, 6, 297-301.

415. Davis, R. C. & Berry, F. Gastrointestinal reactions during a noise avoidance task. Psych. Rep., 1963, 12, 135-137.

Some of the techniques used by Brady (1958) to produce peptic ulcers in monkeys were adapted for use with human subjects. In Brady's experiment the "executive" monkey who pressed a lever to avoid an electric shock developed ulcers, while the control monkey who received the shocks with no means of avoiding them did not. In the present experiment, a noise avoidance task was given to the experimental subject, while the control subject was presented with the same noise stimuli but was unaware of the avoidance task for the other subject.

416. Davis, R. C., Garafolo, L. & Gault, F. P. An exploration of abdominal potentials. J. Comp. Physiol. Psychol., 1957, 50, 519-523.

A technique for recording electrical potentials originating in the abdomen from electrodes on the abdominal surface is described. The slow periodic and aperiodic waves recorded are described and the possible role of other bodily signal generators, respiration, bodily movements, heart, the galvanic skin response, and general tissue potentials is evaluated. It is concluded that abdominal events in the stomach, intestines, and possibly other organs produce many of the observed waves.

417. Davis, R. C., Garafolo, L. & Kveim, K. Conditions associated with gastrointestinal activity. J. Comp. Physiol. Psychol., 1959, 52, 466-475.

A series of experiments investigated the effects of food, rest, visual stimulation, and a gastric balloon on activity recorded from the abdomen by external electrodes and compared this electrical activity with that recorded from a gastric balloon. The correspondence between the electrical and mechanical records is sometimes only general and sometimes quite exact. These recordings are evidently specific to locality. Under resting conditions the recorded activity of the stomach, whether empty or not, declines but is at its lowest when the stomach is empty. The Cannon-Carlson "hunger contractions" are evidently rather rare in such conditions. Activity can be induced by a strong visual stimulus. Introducing a balloon into the stomach increases the electrical activity of the stomach. This increase, it is suspected, is a principal factor in producing the Cannon-Carlson waves. In contrast to some previous results our subjects were rarely successful in reporting these waves even when they were present.

418. Davison, J. S. Letter: Gastric emptying in labour. Lancet, 1975, 2, 227-228.

419. De Bonis, V. & Midulla, L. Influenza della temperatura sui movimenti dello stomaco di Rana. Arch. di fisio., Firenze, 1911, 9, 367-376.

420. De, H. & Lobo, E. Congenital pyloric stenosis: Its clinical variations. Practitioner, 1968, 201, 354-356.

421. Deaton, W. R., Jr. & Bradshaw, H. H. Treatment of vagotomy-induced chronic pylorospasms in dogs. Gastroenterology, 1951, 18, 613-617.

422. Debas, H. T., Farooq, O. & Grossman, M. I. Inhibition of gastric emptying is a physiological action of cholecystokinin. Gastroenterology, 1975, 68, 1211-1217.

423. Deckner, C. W., Hill, J. T. & Bourne, J. R. Shaping of gastric motility in humans. Proc. Am. Psychol. Assoc., 1972, 2, 759-760.

The successful training of three of four nonclinical subjects to reliably increase or decrease their gastric slow-wave activity as measured using cutaneous electrodes is reported. Subjects were trained in 5 sessions, each consisting of 24 minutes of biofeedback in the form of visual inspection of an electrogastrogram record and an auditory analog to the EGG wave amplitude.

424. Dedlovskaia, V. I. Conditions of development of a negative pressure in the stomach and duodenum (radiotelemetric data). Biull. Eksp. Biol. Med., 1973, 76, 44-47.

425. Delestre, J. P. & Godlewski, J. L. Radiologic study of gastric transit after ingestion of natural soda water. Semaine Hop. Paris, 1975, 51, 303-309.

426. Demling, L. The motility of the gastrointestinal tract. Digestion, 1969, 2, 362-368.

427. Demling, L. Physiology of the epigastrium. Dtsch. Med. Wochenschr., 1974, 99, 1029-1032.

428. Demling, L. & Ottenjann, R. Gastrointestinal Motility. International Symposium on Motility of the GI-Tract at Erlangen, 1969. New York: Academic Press, 1971.

429. Dewey, M. M. The anatomical basis of propagation in smooth muscle. Gastroenterology, 1965, 49, 395-402.

430. Diamant, N. E. & Akin, A. N. Response of the lower esophageal sphincter to gastric contractile activity. Gastroenterology, 1970, 58, 940.

431. Diamant, N. E. & Akin, A. N. Effect of gastric contractions on lower esophageal sphincter. Gastroenterology, 1972, 63, 38-44.

432. Diamond, J. & Brody, M. Effect of catecholamines on smooth muscle motility and phosphorylase activity. J. Pharmacol. Exp. Therap., 1966, 152, 202-211.

433. Diaz, Y., Diaz, J., Ramirez, S. E., Shorpinsker, F. & Karchmer, S. Study of fetal gastrointestinal transit in normal pregnancy. Ginecol. Obstet. Mex., 1971, 29, 597-607.

434. Diefenbach, C. O. Gastric function in Caiman crocodilus (Crocodylia: Raptilia) I. Rate of gastric digestion and gastric motility as a function of temperature. Comp. Biochem. Physiol., 1975, 51, 259-265.

435. Dimson, S. B. Transit time related to clinical findings in children with recurrent abdominal pain. Pediatrics, 1971, 47, 666-674.

436. Disebaeva, L. G. & Khorunzhi, I. G. V. Gastric motor function in newborn infants. Vopr. Okhr. Materin. Det., 1975, 20, 53-54.

437. Dixon, W. E. The innervation of the frog´s stomach. J. Physiol., London, 1902, 28, 55-75.

438. Dodds, D. C., Ould, C. L. & Dailey, M. E. The effect of tetraethylammonium chloride on gastric motility in man. Gastroenterology, 1948, 10, 1007-1009.

439. Donovan, I. A., Clarke, R. J., Gunn, I. F. & Alexander-Williams, J. A comparison of gastric emptying at 3 and 12 months after proximal gastric or selective vagotomy without pyloroplasty. Br. J. Surg., 1974, 61, 889-892.

440. Donovan, I. A., Griffin, D. W., Harding, L. K. & Alexander-Williams, J. Paradoxical gastric emptying after gastric surgery in man. Br. J. Surg., 1974, 61, 916-917.

441. Donovan, I. A., Gunn, I. F., Brown, A. & Alexander-Williams, J. Proceedings: Gastric emptying before and after vagotomy and pyloroplasty and vagotomy and antrectomy. Gut, 1973, 14, 824-825.

442. Doret, J. P. Contribution al etude de la motricite normale et pathologique de l´estomac analysee. Gastroenterologia, (Basel), 1951, 77, 231-281.

443. Dougherty, R. W., Riley, J. L. & Cook, H. M. Changes in motility and pH in the digestive tract of experimentally overfed sheep. Am. J. Vet. Res., 1975, 36, 827-829.

444. Douglas, D. M. & Mann, F. C. The gastro-ileal reflex: Further experimental observations. Am. J. Dig. Dis., 1940, 7, 53-57.

445. Doyon, M. Sur l´inhibition du tonus et des mouvements de l´estomac chez le chien. Arch. de physiol. norm. et path., Paris, 1895, 7, 374-384.

446. Dozois, R. R. & Kelly, K. A. Effect of gastrin pentapeptide on gastric emptying of liquids. Surg. Forum, 1970, 21, 293-295.

447. Dozois, R. R. & Kelly, K. A. Effects of a gastrin pentapeptide on canine gastric emptying of liquids. Am. J. Physiol., 1971, 221, 113.

448. Dozois, R. R., Kelly, K. A. & Code, C. F. Effects of distal antrectomy on gastric emptying of liquids and solids. Gastroenterology, 1971, 61, 675.

449. Dragstedt, L. R. Gastric vagotomy in the treatment of peptic ulcer. Postgrad. Med. J., 1951, 10, 482.

450. Dragstedt, L. R., Oberhelman, H. A., Evans, S. O. & Rigler, S. P. Antrum hyperfunction and gastric ulcer. Ann. Surg., 1954, 140, 396-404.

451. Dragstedt, L. R. & Woodward, E. R. Appraisal of vagotomy for peptic ulcer after 7 years. J. Am. Med. Assoc., 1951, 145, 795.

452. Dragstedt, L. R. & Woodward, E. R. Gastric stasis, a cause of gastric ulcer. Scand. J. Gastroenterol., 1970, 6, 243.

453. Dragstedt, L. R., Woodward, E. R., Harper, P. V. & Storer, E. H. Mechanism of relief of ulcer distress by gastric vagotomy. Gastroenterology, 1948, 10, 200-204.

454. Dreyer, N. B., Ray, L. & Larson, V. The action of barbiturates on the motility of the cat´s stomach and intestines. Fed. Proc., 1944, 3, 70.

455. Dubich, I. A. Methods of simultaneous registration of gastric juice secretion and gastric motor activity. Fiziol. Zh. S.S.S.R., 1969, 55, 647-649.

456. Dubois, A. & Gardner, J. Letter: Gastric emptying. New Engl. J. Med., 1974, 291, 104-105.

457. Dubois, A., Gardner, J. & Natelson B. H. Kinetics of monkey gastric function. Proceedings of the Fifth World Congress of Gastroenterology, Mexico City, 1974.

458. Dubois, A., Henry, D. P. & Kopin, I. J. Plasma catecholamines and postoperative gastric emptying and small intestinal propulsion in the rat. Gastroenterology, 1975, 68, 466-469.

459. Duke, G. E., Djuk, H. E. & Evanson, O. A. Gastric pressure and smooth muscle electrical potential changes in turkeys. Am. J. Physiol., 1972, 222, 167-173.

460. Duke, G. E., Evanson, O. A., Ciganek, J. G., Miskowiec, J. F. & Kostuch, T. E. Inhibition of gastric motility in turkeys by intraduodenal injections of amino acid solutions. Poult. Sci., 1973, 52, 1749-1756.

461. Dumas, J. Action mechanism of gastric emptying inhibition in white rats. II. Intervention of the pneumogastric nerve. C. R. Soc. Biol., Paris, 1973, 167, 1843-1846.

462. Dumas, J., Merdrignac, T. & Peres, G. Action of ultraviolet irradiation on gastric transit: Study by 2 different methods. C. R. Soc. Biol., Paris, 1970, 164, 1735-1738.

463. Dumas, J. & Peres, G. The mechanism of action of restraint on gastric emptying in the rat. Role of the adrenal glands. C. R. Soc. Biol., Paris, 1974, 168, 47-50.

464. Duncan, D. L. The effects of vagotomy and splanchnicotomy on gastric motility in the sheep. J. Physiol., London, 1953, 119, 157-169.

465. Dupont, J. R. & Sprinz, H. The neurovegetative periphery of the gut. A reevaluation with conventional techniques in the light of modern knowledge. Am. J. Anat., 1964, 114, 393-402.

466. Duret, R. L., Mainguet, P., Thys, O. & Arguello, M. Disorders of gastrointestinal motility in organic pathology of the digestive tract. Acta Gastroenterol. Belg., 1967, 30, 669-694.

467. Dussardier, M. Origine des contractions rythmiques de l'estomac chez les ruminants. J. Physiol., Paris, 1959, 51, 457-458.

468. Duthie, H. L., Kwong, N. K., Brown, B. H. & Whittaker, G. E. Pacesetter potential of the human gastroduodenal junction. Gut, 1971, 12, 250-256.

469. Dzink, H. E. Radiotelemetry for stomach motility studies in ruminants. Med. Electr. Biol. Eng., 1964, 2, 281.

470. Edwards, A. E., Hill, R. A. & Treadwell, T. Improvements in the magnetometer technique of measuring gastric motility. Psychophysiology, 1967, 4, 116-118.

471. Edwards, A. E. & Treadwell, T. Comparative effects of anger and epinephrine upon stomach motility and the cardiovascular system. Proc., 77th Annual Convention, American Psychological Association, 1969.

472. Edwards, D. A. W. Techniques and interpretations of measurement of gastrointestinal pressures. Ann. Roy Coll. Surg., 1965, 37, 275-284.

473. Edwards, D. A. W. & Rowlands, E. N. An air-filled metal capsule optical manometer system for measuring pressures in the human alimentary canal. In C. N. Smythe (Ed.), Medical Electronics. London: Iliffe, 1960.

474. Edwards, D. A. W. & Rowlands, E. N. Physiology of the gastroduodenal junction. In C. F. Code (Ed.), Handbook of Physiology. Sect. 6, Alimentary Canal. Vol. IV. Motility. Washington, D.C.: American Physiological Society, 1968.

475. Ehrlein, H. J. & von Englehardt, W. The gastric motility of the llama. Zentralbl. Veterinaermed., 1971, [A], 18, 181-191.

476. Eisner, M. Effect of metoclopramide on gastrointestinal motility in man. A monometric study. Am. J. Dig. Dis., 1971, 16, 409-419.

477. Eldh, J., Kewenter, J., Kock, N. G. & Olson, P. Long-term results of surgical treatment for dumping after partial gastrectomy. Br. J. Surg., 1974, 61, 90-93.

478. Eley, K. G., Bennett, A. & Scholes, G. B. A possible role for prostaglandins in gastrointestinal motility. Am. J. Dig. Dis., 1971, 16, 552.

479. Elias, E., Gibson, G. J., Greenwood, L. F., Hunt, J. N. & Tripp, J. H. The slowing of gastric emptying by monosaccharides in test meals. J. Physiol., London, 1968, 194, 317-326.

480. Eliasson, S. G. Cerebral influences on gastric motility in the cat. Acta Physiol. Scand., 1952, 26, 1-70.

481. Eliasson, S. G. Activation of gastric motility from the brain stem of the cat. Acta Physiol. Scand., 1954, 30, 199-214.

482. Eliasson, S. G. Central control of digestive function. In J. Field, H. W. Magoun, and V. E. Hall (Eds.), Handbook of Neurophysiology, Sect. 1, Vol. II. Washington, D.C.: American Physiological Society, 1960.

483. Ellis, F. H., Jr. Surgical management of motility disturbances of the esophagus and cardia. Del. Med. J., 1971, 43, 61-67.

484. Enderle, E. & Pallova, S. Effect of pentagastrin preparations on motor activity of gastric angulus and parietal secretion. Cas. Lek. Cesk., 1971, 110, 637-638.

485. Enderle, E. & Pallova, S. Two types of gastric motor response to atropine administered by the intravenous routed, Cesk. Gastroenterol. Vyz., 1974, 28, 545-550.

486. Engel, B. T. & McFall, R. A. The magnetometer method for recording localized movements of the gastric wall. J. Appl. Physiol., 1959, 14, 1069-1070.

487. Engelholm, L., Govaerts, J. P., Kiekens, R. & Potvliege, R. Gastro-intestinal transit with a test meal in patients having undergone surgery of the stomach. Radiocinematographic study. Med. Audiov., Paris, 1966, 5, 191-200.

488. Esenin, V. I. State of the motor-evacuatory function of the stomach and duodenum in ulcer disease. Sov. Med., 1974, 37, 40-46.

489. Esenin, V. I. & Arablinskii, V. M. Combined radiotelemetric and radiographic study of the motorevacuatory function of the stomach and duodenum. Sov. Med., 1970, 33, 13-15.

490. Fang, H. S. & Chen, C. F. The influence of long-term intermittent exposures to hypoxia on gastric emptying time in rats. Space Life Sci., 1973, 4, 335-337.

491. Farr, C. B., Leuders, C. W. & Bond, E. D. Studies of gastric secretion and motility in mental patients. Am. J. Psychiat., 1925, 5, 93-101.

492. Farrar, J. T. Gastrointestinal smooth muscle function. Am. J. Dig. Dis., 1963, 8, 103-110.

493. Farrar, J. T. Electronics in gastroenterology. Ann. N.Y. Acad. Sci., 1964, 118, 53-64.

494. Farrar, J. T. & Berstein, J. S. Recording of intraluminal gastrointestinal pressures by a radiotelemetering capsule. Gastroenterology, 1958, 35, 603-612.

495. Farrar, J. T. & Bernstein, J. S. Recording of intraluminal gastrointestinal pressures by a radiotelemetering capsule. Gastroenterology, 1968, 54, suppl., 770-771.

496. Farrar, J. T. & Davidson, M. Measurement of gastrointestinal motility in man. In H. Brunner (Ed.), Methods Med. Res., Vol. 8. Chicago: Year Book Publishers, 1960.

497. Farrar, J. T. & Ingelfinger, F. J. Gastrointestinal motility as revealed by study of abdominal sounds. Gastroenterology, 1955, 29, 789-802.

498. Farrar, J. T., Zworykin, V. K. & Baum, J. Pressure-sensitive telemetering capsule for study of gastrointestinal motility. Science, 1957, 126, 975-976.

499. Farrell, J. I. & Ivy, A. C. Studies on the motility of the transplanted gastric pouch. Am. J. Physiol., 1926, 76, 227.

500. Fasth, S., Hulten, L., Jahnberg, T. & Martinson, J. Comparative studies on the effects of bardykinin and vagal stimulation on motility in the stomach and colon. Acta Physiol. Scand., 1975, 93, 77-84.

501. Fauley, G. B., Strauss, H. A. & Ivy, A. C. Studies on the effects of subtotal gastric resection in the dog. Am. J. Surg., 1932, 17, 427-433.

502. Faulhaber, M. & von Redwitz, E. F. Ueber den Einfluss der "circularen Magenresektion" auf die Sekretion und Motilitat des Magens. Med. Klin., 1914, 10, 680-684.

503. Feaver, B. D., Crispin, I. S. & Lind, J. F. Postoperative gastric motility. Can. J. Surg., 1969, 12, 410-414.

504. Fedor, J. H. & Russell, R. W. Gastrointestinal reactions to response-contingent stimulation. Psych. Rep., 1965, 16, 95-113.

Research by Brady and his collaborators (1958) and by Davis and Berry (1963) has shown that the gastrointestinal system reacts quite differently to response-contingent aversive stimulation than it does when the stimulation is not-response-contingent. The present experiment was designed to analyze in greater detail the system's reactions to the first of these two stimulus conditions. Nineteen student volunteers served as subjects in a research design based upon the use of each S as his own control. Continuous electrophysiological recordings of gastrointestinal activity were taken from surface electrodes on the abdomen of each S during the initial 15-min. rest period and an immediately subsequent 25-min. task period. During the task period S could avoid an aversive auditory stimulus if he pressed a key at the proper time intervals; the task was sufficiently difficult that successful avoidances for different subjects ranged from an over-all 16% to 72% with a mean of 44% for the group as a whole. The resulting data made it possible to analyze the differential reactions of the gastrointestinal system on successful and unsuccessful trials when the occurrence of the aversive stimulation was always response-contingent, i.e., under the potential control of S. Three parameters of the electrophysiological manifestations of gastrointestinal activity were studied: amplitude, displacement, and peak response time. Analyses of the results show that the mean amplitude of gastrointestinal activity increases significantly only when avoidance of the aversive stimulus was unsuccessful; during successful trials the amplitude remained at pre-task baseline levels. This differential reaction was strikingly apparent when the outcome of S's response differed on successive trials: when both trials of a pair involved the same response contingencies, i.e., either successful or unsuccessful avoidance, gastrointestinal reactions were not significantly different; but large and very significant changes occured when pairs of trials involved different response contingencies. In contrast to the effects obtained by Davis and Berry (1963) under non-response-contingent conditions when very rapid adaptation occurred, our results show a significant trend toward increasing amplitudes of gastrointestinal activity as the task trials continued. Throughout the entire task period peak response times on unsuccessful trials were significantly shorter than on successful trials. Similar analyses of the displacement measure failed to show any significant effects. The results of the present experiment when considered in relation to those provided by earlier studies suggest that the differences in effects of response-contingent and of non-response-contingent conditions cannot be accounted for in terms of the physical characteristics of the aversive stimulus, of sensory processes per se, or of motor output. There is no support for the possibility that there may occur some interaction between level of performance of the response-contingent task and gastrointestinal activity or that the number of aversive reinforcements during a contingency period is a critical factor. There remains the possibility that the critical factors involved are central, rather than peripheral in nature.

505. Feldman, S. & Gilbaldi, M. Effect of bile salts on gastric emptying and intestinal transit in the rat. Gastroenterology, 1968, 54, 918-921.

506. Feldman, M. & Morrison, S. Effect of insulin on motility of the stomach following bilateral vagotomy. Am. J. Dig. Dis., 1948, 15, 175.

507. Feldman, M. & Morrison, S. Experimental study of gastric emptying in vagotimized dog. Gastroenterology, 1948, 10, 502-503.

508. Feng, C. I., Chin, Y., Wang, N. C. & Chang, S. S. The pharmacology of sinomenine. VII. Effect of sinomenine on the gastro-intestinal movement and its mechanism. Acta Pharm. Sin., 1965, 12, 492-495.

509. Fennegan, F. M. & Puiggari, M. J. Hypothalamic and amygdaloid influence on gastric motility in dogs. J. Neurosurg., 1966, 24, 497-504.

510. Fenton, P. F. & Dickson, H. M. Changes in some gastrointestinal functions following x-irradiation. Am. J. Physiol., 1954, 177, 528-530.

511. Fenton, P. F. & Pierce, H. B. Response of the gastrointestinal tract to ingested glucose. Fed. Proc., 1944, 3, 57.

512. Fenton, P. F. & Pierce, H. B. Response of gastrointestinal tract to ingested cereal starch. Am. J. Physiol., 1947, 148, 296-301.

513. Fetter, D., Barron, L. & Carlson, A. J. The effect of induced hyperthyroidism on the gastrointestinal motility of vagotomized dogs. Am. J. Physiol., 1932, 101, 605-611.

514. Fetter, D. & Carlson, A. J. Effects of experimental hyperthyroidism on gastro-intestinal motility. Am. J. Physiol., 1932, 101, 598-604.

515. Fisher, R. S. & Cohen, S. Pyloric-sphincter dysfunction in patients with gastric ulcer. New Engl. J. Med., 1973, 288, 273-276.

516. Fishlock, D. J., Parks, A. G. & Dewell, J. V. Action of t-hydroxytryptamine on the human stomach, duodenum, and jejunum in vitro. Gut, 1965, 6, 338-342.

517. Fjuii, I. & Mizonishi, T. Proceedings: Electric activity of the gastric motility center of the medulla oblongata and its relation to reflex control by the splanchnic nerves. J. Physiol. Soc. Jap., 1974, 36, 289.

518. Fleischer, K. & Sewing, K. Fr. Die Motilitat Von Antrum, Pylorus Und Duddenum Beim Kaninchen. Naunyn-Schmiedebergs Arch. Pharmakol. Exp. Patholog., 1968, 260, 275-283.

519. Flint, F. J. & Grech, P. Pyloric regurgitation and gastric ulcer. Gut, 1970, 11, 735-737.

520. Folkow, B. & Rubinstein, H. Behavioral and gastrointestinal changes (motility and blood-flow) induced by electrical stimulation of the lateral hypothalamus in cats. Acta Physiol. Scand., 1963, 213, 44.

521. Fontana, G., Azzaroli, P., Lanfranchi, G. A. & Vezzadini, P. Description of a method for evaluation of the gastro-duodeno-jejunoileal transit time of a test meal. Minerva Gastroenterol., 1968, 14, 205-213.

522. Frankel, L. A. The pump mechanism of the stomach and gastric evacuation. I. Int. Surg., 1968, 50, 17-33.

523. Frankel, L. A. The pump mechanism of the stomach and gastric evacuation. II. Int. Surg., 1968, 50, 155-163.

524. Frankendal, B. Gastric emptying and small intestinal propulsion in mice following irradiation of the abdomen. Acta Radiol., [Ther.] (Stockh.), 1973, 12, 529-540.

525. Frankendal, B. & Junghagen, P. Gastrointestinal transit time in ovarian carcinoma irradiated with 60CO. Acta Radiol., [Ther.] (Stockh.), 1974, 13, 129-140.

526. French, J. D., Porter, R. W., Cavanaugh, E. B. & Longmire, R. L. Experimental gastroduodenal lesions induced by stimulation of the brain. Psychosom. Med., 1957, 19, 209-220.

527. Freytes, M. A. & Amuchastegui, L. M. Gastric motility in chronic chagasics. Prensa. Med. Argent., 1968, 55, 1239-1243.

528. Fukuhara, T. Current trends in studies of gastrointestinal motility. Jap. J. Smooth Muscle Res., 1968, 4, 1-14.

529. Fukushima, K., Okada, I. & Nakasono, Y. Studies on the electrogastrograph. Med. J. Osaka Univ., 1951, 2, 83.

530. Furness, J. B. & Costa, M. The adrenergic innervation of the gastrointestinal tract. Ergeb. Physiol., 1974, 69, 2-51.

531. Fyke, F. E., Jr., Code, C. F. & Schlegel, J. F. Gastroesophageal sphincter in healthy human beings. Gastroenterologia, (Basel), 1956, 86, 135-150.

532. Gahagan, T. The function of the musculature of the esophagus and stomach in the esophagastric sphincter mechanism. Surg. Gynecol. Obstet., 1962, 114, 293-303.

533. Gal, A. N. & Navarro, J. L. Electrosplanchnography (E.S.G.) and electrogastroenterography. Rev. Esp. Enferm. Apar. Dig., 1972, 36, 587-592.

534. Galitskaia, N. A. The role of posterior radicular innervation in the motor function of the stomach during ontogenesis in young dogs. Fiziol. Zh. S.S.S.R., 1965, 51, 990-996.

535. Gallia, F. J. Diffuse, gastrointestinal tract motor abnormality. J. Am. Med. Assoc., 1968, 205, 235-237.

536. Garafolo, L. & Davis, R. C. Gastrointestinal activity in hunger and after food: The question of hunger pangs. Proc. Indiana Acad. Sci., 1957, 67, 298.

By a new method the activity of the stomach and intestinal tract was recorded without the insertion of any foreign object into the stomach or other parts. Records were made on 8 subjects after at least eight hours without food and after their consumption of a light breakfast. Records were measured so the amplitude and frequency of waves may be compared. Characteristic rhythms of stomach and intestinal tract are observed. These are generally quite small or absent before the food and increase significantly in all four quadrants of the abdomen after eating. There is a significant intercorrelation in the increase in the several locations. The results cast doubt on the common belief that during hunger there are pangs associated with large slow contractions of the stomach. These may be induced by the balloon introduced into the stomach as a recording device.

537. Garrett, J. M., Summerskill, W. H. & Code, C. F. Antral motility in patients with gastric ulcer. Am. J. Dig. Dis., 1966, 11, 780-789.

538. Gaskell, W. H. On the structure, distribution and function of the nerves which innervate the visceral and vascular systems. J. Physiol., London, 1886, 7, 18-80.

539. Gaudeau, C., Thouvenot, J. & Martin, A. Techniques de reconnaissance des processus dynamigues dans l'electrosplanchnogramme (ESG). Gen. Biol. Med., 1970, 1, 44-55.

540. Gaussen, M. A. L'Auscultation de l'abdomen. J. Mid. Bordeaux, 1952, 129, 25.

541. Gavend, M. & Gavend, M. Study in anesthetized dogs of the effects of the pentazocine or the tonus of Oddi's sphincter, gastro-intestinal motility and the activity of the ureter. Therapie, 1970, 25, 785-794.

542. Geall, M. G., Code, C. F., McIlrath, D. C. & Summerskill, W. H. Measurement of gastrointestinal transmural electric potential difference in man. Gut, 1970, 11, 34-37.

543. Geall, M. G., Phillips, S. F. & Summerskill, W. H. Profile of gastric potential difference in man. Effects of aspirin, alcohol, bile, and endogenous acid. Gastroenterology, 1970, 58, 437-443.

544. Geiger, E. & Pinsky, J. J. Effect of changes in environmental temperature on stomach emptying in rats. Proc. Soc. Exp. Biol. Med., 1956, 91 107-110.

545. Geisel, A., Arai, T., Jefferson, N. C. & Necheles, H. Gastric motility in the dog: Distinctive activity of the subcardiac zone of the stomach. Am. J. Gastroenterol., 1965, 44, 551-558.

546. Genes, S. G. & Lesnoi, N. G. Evacuatory gastric function in adrenolectomy and under the influence of a glucocorticoid surplus. Fiziol. Zh. S.S.S.R., 1969, 55, 1272-1277.

547. George, J. D. Gastric acidity and motility. Am. J. Dig. Dis., 1968, 13, 376-383.

548. George, J. D. New clinical method for measuring the rate of emptying: The double sampling test meal. Gut, 1968, 9, 237-242.

549. George, J. D., Connell, A. M. & Kennedy, T. Gastric emptying patients with post-vagotomy diarrhea. Gut, 1968, 9, 732.

550. Gerasimenko, A. A., Khoromski, L. N. & Kit, O. N. State of the motor-evacuation function of the stomach before and after gastric resection in patients with ulcer disease. Klin. Khir., 1975, 40-42.

551. Gerasimets, M. T. Comparative assessment of the motor activity of the stomach stump and the small intestine after reconstructive gastroduodenojejunoplasty. Eksp. Khir. Anesteziol., 1968, 13, 22-24.

552. Gershon-Cohen, R., Shay, H. & Fels, S. S. The relation of meal temperature to gastric motility and secretion. Am. J. Roentgenol. Radium Ther. Nucl. Med., 1940, 43, 237-242.

553. Gershon, M. D. Inhibition of gastrointestinal movement by sympathetic nerve stimulation: The site of action. J. Physiol., London, 1967, 189, 317-327.

554. Gershon, M. D. Serotonin and the motility of the gastrointestinal tract. Gastroenterology, 1968, 54, 453-456.

555. Gianturco, C. Some mechanical factors of gastric physiology. Am. J. Roentgen., 1934, 31, 735-744.

556. Giles, G. R., Humphries, C., Mason, M. C. & Clark, C. G. Effect of pH changes on the cardiac sphincter. Gut, 1969, 10, 852-856.

557. Giles, G. R., Mason, M. C., Humphries, C. & Clark, C. G. Action of gastrin on the lower oesophageal sphincter in man. Gut, 1969, 10, 730-734.

558. Gillespie, J. S. & MacKenna, B. R. The inhibitory action of the sympathetic nerves on the smooth muscle of the rabbit gut, its reversal by reserpine and restoration by catechol amines and by DOPA. J. Physiol., London, 1961, 156, 17-34.

559. Ginsberg, R. S., Feldman, M. & Necheles, H. Effect of odors on appetite. Gastroenterology, 1948, 10, 281-285.

560. Ginsburg, H. & Tumpowsky, I. Contributions to the physiology of the stomach. L. Studies on the control of hunger by drugs. Arch. Int. Med., 1918, 22, 553-570.

561. Ginzel, K. H. Are mucoscal nerve fibres essential for the peristaltic reflex? Nature, 1959, 184, 1235-1236.

562. Giuren-Benetato, A. M. Disorders of motility of the digestive tract. Stud. Cercet. Fiziol., 1968, 13, 203-216.

563. Gladilov, N. Effect of merichlersk mineral water treatment on the motor-evacuatory function of the stomach in patients with peptic ulcer. Vopr. Kurortol. Fizioter. Lech. Fiz. Kult., 1968, 33, 431-435.

564. Glaessner, K. Beitrag zur Kenntniss der Magenbewegungen. Pfluger's Arch. f. d. gesammt. Physiol., 1901, 86, 291-306.

565. Glatzel, H. & Hackenberg, K. Radiologic studies on the effect of capscium spice and mustard on the motility and secretion of the digestive tract. Dtsch. Z. Verdav. Stoffwechsselkr., 1966, 26, 113-121.

566. Glatzel, H. & Hackenberg, K. Effects of caffeine containing and decaffeinated coffee on the digestive functions X-ray studies of the secretion and peristalsis of stomach, intestines and gallbladder. Med. Klin., 1967, 62, 625-628.

567. Glavtcheva, L., Manchanda, S. K., Box, B. & Stevenson, J. A. F. Gastric motor activity during feeding induced by stimulation of the lateral hypothalamus in the rat. Can. J. Physiol. Pharmacol., 1972, 50, 1091-1098.

568. Glavtcheva, L., Rozkowska, E. & Fonberg, E. The effect of lateral hypothalamic lesions on gastric motility in dogs. Acta Neurobiol. Exp. (WARSZ), 1970, 30, 279-293.

569. Glavtcheva, L. Studies concerning the mechanism of action of largactil on gastric motility. Rev. Roum. Physiol., 1968, 5, 51-62.

570. Glavtcheva, L. & Velkova, V. Studies of the functional division of the hypothalmus in relation to its effects on gastric motility. Rev. Roum. Physiol., 1971, 8, 73-80.

571. Gloor, M. & Heinkel, K. Measurement of the emptying time of the stomach by means of the alkali test. Stomach function tests with the endoradio probe. Med. Welt., 1969, 47, 2575-2579.

572. Gocke, C. Beitrage zur Morphologie des Magens nach Resektionen. Beitr. z. klin. Chir., 1916, 99, 294-382.

573. Goetze, O. Die Motilitat und Sekretion des operierten Magens. In Bethe, Bergmann, Embden, and Ellinger (Eds.), Handbuch der Normalen und Pathologischen Physiologie. Berlin: Julius Springer, 1927.

574. Goetze, O. Die Mechanik des gesunden und kranken Magens. Arch. f. klin. Chir., 1929, 157, 601-605.

575. Golenhofen, K. Intrinsic rhythms of gastrointestinal tract. In L. Demling & R. Ottenjann (Eds.), Gastrointestinal Motility. New York: Academic Press, 1971.

576. Golenhofen, K., von Loh, D. & Milenov, K. Spontaneous activity of isolated smooth muscle preparation from different regions of the guinea pig stomach. Pfluger's Arch. f. d. gesammt. Physiol., 1970, 315, 336-356.

577. Goltz, F. Studien uber die Bewegungen der Speiserohre und des Magens des Frosche. Pfluger's Arch. f. d. gesammt. Physiol., 1872, 6, 616-642.

578. Gomez-Perez, F. J., Ryan, J. R. & Staub, R. Influence of phenformin on gastric emptying rate. J. Clin. Pharmacol., 1974, 5, 261-263.

579. Gomez-Perez, F. J., Ryan, J. R., Staub, R. & Vargas, J. R. Gastric emptying rate as measured by a radioisotopic technique. J. Med. Assoc. State Ala., 1974, 43, 493-495.

580. Goodman, E. N. Improved method of measuring the potential difference across the human gastric membranes and its clinical significance. Surg. Gynecol. Obstet., 1942, 75, 583-592.

581. Goodman, E. N., Colcher, H., Katz, G. M. & Dangler, C. L. The clinical significance of the electrogastrogram. Gastroenterology, 1955, 29, 598-607.

582. Goodman, E. N., Ginsberg, I. A. & Robinson, M. A. An improved apparatus for measuring the electrogastrogram. Science, 1951, 113, 682.

583. Goodman, E. N., Sandler, B. & Sullivan, M. R. A probe for measuring the electrical activity of the human stomach. Am. J. Gastroenterol., 1964, 42, 511-514.

584. Goodman, E. N., Sandler, B. & Sullivan, M. R. An endoscopic method for recording DC potentials in the upper gastrointestinal tract. Am. J. Gastroenterol., 1975, 63, 13-19.

585. Goodman, E. N. & Schlaeger, R. Electrogastrographic and radiologic studies in patients postgastrectomy. Bull. N.Y. Acad. Med., 1959, 35, 765.

586. Goodman, R. D., Lewis, A. E. & Schuck, E. A. Effects of x-irradiation on gastrointestinal transit and absorption availability. Am. J. Physiol., 1952, 169, 242-247.

587. Goodman, R. D., Lewis, A. E., Schuck, E. A. & Greenfield, M. A. Gastrointestinal transit. Am. J. Physiol., 1952, 169, 236-241.

588. Gorskaia, A. A., Gracheva, L. C. & Usievich, M. A. Materialy K voprosu mekhanizme vozniknoveniia i protesse protekaniia periodicheskoi deiatel′nosti zheludka (Data on the periodic activity of the stomach and processes of its course). Fiziol. Zh. S.S.S.R., 1949, 35, 647-652.

589. Grahame, G. R., Garrett, J. M. & Hirschowitz, B. I. Two-deoxyglucose and histamine effects on gastric motility and secretion in the dog. Am. J. Physiol., 1968, 215, 243-248.

590. Grechishkin, L. L. The significance of stem portions of the brain in the action mechanism of the central pharmacological agents on periodic contractions of the stomach. Boll. Exp. Biol. Med., 1963, 4, 52-56.

591. Grechishkin, L. L. The effect of central neurotropic agents on the periodic contraction of the dog's stomach, the agents being administered into various arteries of the brain. Boll. Exp. Biol. Med., 1964, 11, 63-66.

592. Greeff, K., Kasperat, H. & Osswald, W. Paradoxical effects of spectrical stimulation of the vagus nerve on the isolated stomach and atrial preparation of guinea pig and its alteration by ganglionic-blocking agents, sympathetytics, reserpine and cocaine. Arch. Exp. Pathol. Pharmakol., 1962, 243, 528-545.

593. Greengard, H. Digestive system. Ann. Rev. Physiol., 1947, 9, 191-224.

594. Gregory, R. A. The digestive system. Ann. Rev. Physiol., 1954, 16, 155-174.

595. Greitz, T. Roetgenologic and morphologic studies of the mucous membrane movements in the stomach of the cat. Acta Radiol., 1964, 2, 225.

596. Grevsten, S., Johansson, H. & Nylander, G. Analysis of quantitative method of evaluation as applied to the propulsive gastrointestinal motility in the rat. Acta Chir. Scand., 1967, 133, 563-567.

597. Grey, E. G. Observations on the postural activity of the stomach. Am. J. Physiol., 1918, 45, 272-285.

598. Griffith, G. H., Owen, G. M., Campbell, H. & Shields, R. Gastric emptying in health and in gastroduodenal disease. Gastroenterology, 1968, 54, 1-7.

599. Griffith, G. H., Owen, G. M., Kirkman, S. & Shields, R. Measurement of rate of gastric emptying using chromium-51. Lancet, 1966, 1, 1244-1245.

600. Griggs, R. C. & Stunkard, A. J. The interpretation of gastric motility. II. Sensitivity and bias in the perception of gastric motility. Arch. Gen. Psychiat., 1964, 11, 82-89.

601. Grigorev, V. G. The electric potential of the stomach in patients with peptic ulcer and its dependence on the amount of free hydrochloric acid. Sov. Med., 1966, 29, 53-58.

602. Grigorev, V. G. On the periodic activity of the stomach and duodenum. Klin. Med., Mosk., 1967, 45, 26-33.

603. Grinberg, A. A., Cherniakevich, S. A. & Neiman, M. I. Motor-evacuatory function of the stomach after vagotomy in combination with economic resection or pyloroplasty in peptic ulcer. Klin. Med., Mosk., 1971, 49, 46-49.

604. Grinberg, A. A., Groisman, S. D. & Cherniakevich, S. A. The motor function of vagotomized stomachs. Khirurgiia, Mosk., 1972, 48, 97-103.

605. Grinevich, V. A. Motor activity of the gastrointestinal tract in patients with appendicitis. Khirurgiia, Mosk., 1971, 47, 55-58.

606. Groedel, F. M. Die peristolische Funktion des Magens im Rontgenbilde. Munchen Med. Wochenschr., 1909, 56, 567.

607. Groedel, F. M. Die Magenbewegungen. Hamburg: L. Grafe & Sillem, 1912.

608. Groedel, F. M. Die Eigenbewegung des Magens im Rontgenbild. Fortschr. a. d. Geb. d. Rontgenstrahlen, 1925, 33, 8-15.

609. Groisman, S. D. & Beheka, A. D. Volumetric roentgenologic method for the study of the stomach evacuation function. Fiziol. Zh. S.S.S.R., 1972, 18, 700-703.

610. Groisman, S. D. & Beheka, A. D. Analysis of gastric evacuation mechanisms. Fiziol. Zh. S.S.S.R., 1974, 60, 805-811.

611. Groisman, S. D. & Kharchenko, N. M. Evacuatory function of the stomach after its transection and anastomosing. Fiziol. Zh. S.S.S.R., 1972, 59, 1257-1261.

612. Grossman, M. I. Digestive system. Ann. Rev. Physiol., 1950, 12, 205-236.

613. Grossman, M. I. The digestive system. Ann. Rev. Physiol., 1963, 25, 165-194.

614. Grossman, M. I. & Stein, J. F., Jr. Vagotomy and hunger producing action of insulin in man. J. Appl. Physiol., 1948, 1, 263-269.

615. Gruber, C. M., Thomas, J. E., Crider, J. O. & Brundage, J. T. The effect of morphine sulphate and of dihydromogrphinone hydrochloride upon the antrum, pyloric sphincter and duodenum in non-anesthesized dogs. J. Pharmacol. Exp. Therap., 1936, 57, 170-178.

616. Gruber, F. New method of studying anatomy and motility of stomach and duodenum. Exptl. Med. Surg., 1944, 2, 1-35.

617. Grupcev, V., Altaparmakov, Iv., Bosijanov, VI. & Papazova, M. Roentgenographic and electrophysiologic examinations of the tonus and mobility of the stomach in duodenal ulcer and pyloric stenosis. Radiol. Diagn., Berl., 1974, 15, 777-780.

618. Grutzner, P. Ein Beitrag zum Mechanismus der Magenverdauung. Arch. f. d. ges. physiol., 1905, 106, 463-522.

619. Guerrin, F., Milbled, G., Merveille, P. & Hequet, B. Effects of metoclopramide on gastric motility in rats. C. R. Soc. Biol., Paris, 1967, 161, 1318-1322.

620. Habel, R. E. A study of the innervation of the ruminant stomach. Cornell Vet., 1956, 46, 555-628.

621. Hadary, G., Sommer, H. H. & Gance, J. The relationship between the curd tension of milk and gastric emptying time in children. J. Dairy Sci., 1943, 26, 259.

622. Haffner, J. F. Pressure responses to cholinergic and adrenergic agents in the fundus, corpus, and antrum of isolated rat and guinea-pig stomachs. Acta Chir. Scand., 1973, 139, 650-655.

623. Haffner, J. F. & Aune, S. Jansson, G., Stadaas, J. Nervous regulation of gastric motility. Tidsskr. Nor. Laegeforen., 1971, 91, 596-601.

624. Hall, W. H., Herb, R. W. & Brady, J. P. Gastric function during hypnosis and hypnotically induced gastrointestinal symptoms. J. Psychosom. Res. 1967, 11, 263.

625. Hall, W. H. & Read, R. C. Effect of vagotomy on gastric emptying. Am. J. Dig. Dis., 1970, 15, 1047-1053.

626. von Haller, A. First lines of physiology. First Am. Ed. Troy, N.Y., 1803, p. 320.

627. Halmos, T., Forgacs, S. & Rosinger, A. Gastroparesis diabeticorum. Orv. Hetil., 1969, 110, 1451-1454.

628. Halmos, T., Forgacs, S. & Rosinger, A. Diabetic gastroparesis. Munchen Med. Wochenschr., 1970, 112, 780-784.

629. Hamberger, B. & Norberg, K. A. Studies on some systems of adrenergic synaptic terminals in the abdominal ganglia of the cat. Acta Physiol. Scand., 1965, 65, 235-242.

630. Hammer, J. M., Visscher, F. & Hill, E. J. Experimental gastrojejunal ulcers produced by reversing duodenum. Arch. Surg., 1953, 67, 23-28.

631. Hancock, B. D., Bowen-Jones, E., Dixon, R., Dymock, I. W. & Cowley, D. J. The effect of metoclopramide on gastric emptying of solid meals. Gut, 1974, 15, 462-467.

632. Hancock, B. D., Bowen-Jones, E., Dixon, R., Testa, T., Dymock, I. W. & Cowley, D. J. Proceedings: The effect of posture on the gastric emptying of solid meals in normal subjects and patients after vagotomy. Br. J. Surg., 1974, 61, 323.

633. Hancock, B. D., Bowen-Jones, E., Dixon, R., Testa, T., Dymock, I. W. & Cowley, D. J. The effect of posture on the gastric emptying of solid meals in normal subjects and patients after vagotomy. Br. J. Surg., 1974, 61, 945-949.

634. Hansky, J. & Connell, A. M. Measurement of gastrointestinal transit using radioactive chromium. Gut, 1962, 3, 187-188.

635. Hanson, J. & Lowy, J. Structure of smooth muscles. Nature, 1957, 180, 906-909.

636. Harada, M., Tenmyo, N., Aburada, M. & Endo, T. Pharmacological studies of gardenine fructus. I. Effect of geniposide and genipin on the biliary excretion, the gastric juice secretion, and the gastric contraction, and other pharmacological actions. J. Pharmacol. Soc. Jap., 1974, 94, 157-162.

637. Hardt, L. L. J. Pain in active pathologic processes in stomach or duodenum. Gastric and duodenal contractions as the direct cause. J. Am. Med. Assoc., 1918, 70, 837-839.

638. Hardt, L. L. J. Studies of the cause of pain in gastric and duodenal ulcers: II. Peristalsis as direct cause of pain in gastric ulcers with achylia and in duodenal ulcers. Arch. Int. Med., 1922, 29, 684-689.

639. Harichaux, P. & Moline, J. The duodeno-gastric reflex in the rat. Conditions of its appearance. C. R. Soc. Biol. Paris, 1966, 160, 1703-1705.

640. Harichaux, P. & Thouvenot, J. On the periodicity of pyloric evacuation; its independence of the pressure distention of the isolated stomach in rats and mice. Compt. Rend. Soc. Biol., 1962, 156, 737-741.

641. Harichaux, P. & Thouvenot, J. Action de l'acetylcholine sur l'evacuation pylorique chez le rat blanc. Compt. Rend. Soc. Biol., 1963, 157, 1277-1281.

642. Harkins, R. W., Longenecker, J. B. & Sarett, H. P. Effect of type and level of dietary fat on gastric emptying time in rats. Fed. Proc., 1963, 22, 376.

643. Harkins, R. W., Longenecker, J. B. & Sarett, H. P. Effect of type and level of dietary fat on gastric emptying time in rats. Fed. Proc., 1963, 22, 376.

644. Harper, A. A. Vagovagal reflex effects on the gastrointestinal tract. Gastroenterology, 1960, 39, 639-641.

645. Harper, A. A., Kidd, C. & Scratcherd, T. Vago-vagal reflex effects on gastric and pancreatic secretion and gastrointestinal motility. J. Physiol., London, 1959, 148, 417-436.

646. Harris, L. D., Kelly, J. E., Jr. & Kramer, P. Relation of the lower esophageal ring to the esophagogastric junction. New Engl. J. Med., 1960, 263, 1232-1235.

647. Harris, S. C., Grossman, M. I. & Ivy, A. C. Role of vagus nerves in inhibition of gastric motility by fat and by intestinal and urinary extracts. Am. J. Physiol., 1947, 148, 338-343.

648. Harrison, J. S. & McSwiney, B. A. The chemical transmitter of motor impulses to the stomach. J. Physiol., London, 1936, 87, 79-96.

649. Harvey, R. F. Hormonal control of gastroinestinal motility. Am. J. Dig. Dis., 1975, 20, 523-539.

650. Harvey, R. F., Mackie, D. B., Brown, N. J., Keeling, D. H. & Davies, W. T. Measurement of gastric emptying time with a gamma camera. Lancet, 1970, 1, 1616-18.

651. Hato, H. Effects of the electric stimulation of the peripheral cut end of the rami communicantes and of the stomach. J. Physiol. Soc. Jap., 1958, 20, 388-396.

652. Hato, H. Effects of the electric stimulation of the peripheral cut end of the spinal ventral and dorsal roots upon the movements of the frog's stomach. J. Physiol. Soc. Jap., 1958, 20, 397-400.

653. Hato, H. Effects of the electric stimulation of the peripheral cut end of the vagosympathetic nerves on the movements of the frog's stomach. J. Physiol. Soc. Jap., 1958, 20, 401-406.

654. Hato, H. On the motor fibres in the gastric movements of the frog's stomach. J. Physiol. Soc. Jap., 1958, 20, 401-406.

655. Hawkins, G. K., Margolin, S. G. & Thompson, J. J. Gastric emptying time: Comparative studies with placebos, prantal and banthine. Gastroenterology, 1953, 24, 193-199.

656. Heading, R. C., Tothill, P., McLoughlin, G. P. & Shearman, D. J. Proceedings: Gastric emptying rate measurement in man: A method for simultaneous study of solid and liquid phases. Gut, 1974, 10, 841.

657. Heading, R. C., Tothill, P., McLoughlin, G. P., Shearman, D. J. & Carter, D. C. Proceedings: Effects of gastric surgery on differential emptying of solid and liquid components of the gastric contents. Br. J. Surg., 1975, 62, 659.

658. Hedblom, C. A. & Cannon, W. B. Some conditions affecting the discharge of food from the stomach. Am. J. Med. Sci., 1909, 138, 504-521.

659. Hegglin, J., Babotai, I. & Largiader, F. The influence of vagotomy on electrogastromyography. Eur. Surg. Res., 1969, 1, 56-63.

660. Hegglin, J., Largiader, F. & Babotai, I. Electromyographic studies on the stomach before and after vagotomy. Helv. Chir. Acta., 1968, 35, 274-283.

661. Hellebrandt, F. A. & Dimmitt, L. Studies on the influence of exercise on the digestive work of the stomach. III. Its effect on the relation between secretory and motor function. Am. J. Physiol., 1934, 107, 355-363.

662. Hellebrandt, F. A. & Tepper, R. H. Studies on the influence of exercise on the digestive work of the stomach. II. Its effect on emptying time. Am. J. Physiol., 1934, 107, 355-363

663. Hellemans, J., Vantrappen, G. & Vandenbroucke, J. The electrophysiologic basis of gastrointestinal motility. Acta Gastroenterol. Belg., 1967, 30, 634-645.

664. Hemingway, A. Physiological changes during motion sickness. Fed. Proc., 1945, 4, 33.

665. Hemmati, A. & Sawade, H. The influence of acid production and emptying time of the stomach on fat resorption. Med. Welt., 1969, 47, 2580-2584.

666. Hennessy, T., Lehane, M. & O'Laoire, S. Proceedings: The significance of myoelectrical activity in the canine stomach. J. Physiol., London, 1974, 236, 19-20.

667. Henry, G. W. Gastrointestinal motor functions in schizophrenia-roetgenologic observations. Am. J. Psychiat., 1927, 7, 135-152.

668. Henschel, A., Keys, A., Sturgeon, A. M. & Taylor, H. L. Influence of test meal composition on gastric emptying in man. Am. J. Physiol., 1947, 149, 107-111.

669. Henschel, A. & Sturgeon, A. M. The effect of semi-starvation on the emptying of the human stomach. Fed. Proc., 1946, 5, 45.

670. Henschel, A., Taylor, H. L. & Keys, A. The gastric emptying time of man at high and normal environmental temperatures, Am. J. Physiol., 1944, 141, 205-208.

671. Hepburn, J. S. Temperature of the gastrointestinal tract--the effect thereon of hot and cold foods and of physical therapeutic agents. Arch. Int. Med., 1933, 52, 603-615.

672. Hermann, G. & Johnson, V. Management of prolonged gastric retention after vagotomy and drainage. Surg. Gynecol. Obstet., 1970, 130, 1044-1048.

673. Hesser, F. H., Langworthy, O. R. & Kolb, L. C. Experimental study of gastric activity released from cortical control. J. Neurophysiol., 1941, 4, 274-283.

674. Hesser, F. H. & Perret, G. E. Studies on gastric motility in the cat. II. Cerebral and infracerebral influence in control, vagectomy and cervical cord preparations. Gastroenterology, 1960, 38, 231-246.

675. Hiatt, J. F. & Kripke, D. F. Ultradian rhythms in waking gastric activity. Psychosom. Med., 1975, 37, 320-325.

676. Hiatt, R. B., Hinderstein, C. S. & Moore, A. Development of neural control of alimentary motor function in the vertebrates. A comparative study. J. Surg. Res., 1968, 8, 286-292.

677. Higgins, J. R., Schoen, A. M. & Knoefel, P. K. Effects of some synthetic antispasmotic agents on human gastric and duodenal functions. Gastroenterology, 1948, 11, 508-518.

678. Hightower, N. C. Comparison of intraluminal pressures and motility patterns in the gastrointestinal tract of human beings when recorded by falloon and direct (electrical transter) pressure system. Fed. Proc., 1952, 11, 69-70.

679. Hightower, N. C. Motility of the alimentary canal of men. In J. A. Leder & G. C. Noeller (Eds.), Disturbances in Gastrointestinal Motility. Springfield, Ill.: Thomas, 1959.

680. Hightower, N. C. Contributions of a gastrointestinal laboratory to the practice of gastroenterology. Am. J. Dig. Dis., 1961, 6, 798-812.

681. Hightower, N. C. The digestive system. Ann. Rev. Physiol., 1962, 24, 109-138.

682. Hightower, N. C. & Code, C. F. The quantitative analysis of antral gastric motility records in normal human beings, with a study of the effects of neostigmine. Proc. Staff Meetings Mayo Clinic, 1950, 25, 697-704.

683. Hightower, N. C., Code, C. F. & Mahler, F. T. Method for the study of gastrointestinal motor activity in human beings. Proc. Staff Meetings Mayo Clinic, 1949, 34, 453-462.

684. Hightower, N. C., Code, C. F., Mahler, F. T. & Morlock, C. G. Effect of prostionime and urecholine on human gastric motility. Fed. Proc., 1949, 8, 75.

685. Hightower, N. C. & Gambill, E. E. Effects of banthine on pain and gastric antral motility in patients with duodenal ulcer. Gastroenterology, 1953, 23, 244-251.

686. Hightower, N. C., Walters, W. & Morlock, C. G. The effects of urethane of beta-methylchlorine chloride (urecholine) on antral gastric motility in man following vagotomy. Proc. Staff Meetings Mayo Clinic, 1950, 25, 1950.

687. Hildes, J. A. & Dunlop, D. L. A method for estimating the rates of gastric secretion and emptying. Can. J. Med. Sci., 1951, 29, 83-89.

688. Hirota, K. Evaluation of the residual stomach, with reference to motor function and nutrition. Shujutsu, 1968, 22, 1037-1046.

689. Hoebel, B. G. Feeding: Neural control of intake. Ann. Rev. Physiol., 1971, 33, 533-568.

690. Hoebel, B. G. & Thompson, R. D. Aversion to lateral hypothalamic stimulation caused by intragastric feeding or obesity. J. Comp. Physiol. Psychol., 1969, 68, 536-543.

691. Hoeffel, J. C., Benoit, D., Fays, J., Gaucher, P. & Moeller, R. Gastric stasis after vagotomy and pyloroplasty. Am. J. Surg., 1971, 122, 365-367.

692. Hoelzel, F. Central factors in hunger. Am. J. Physiol., 1927, 82, 665-671.

693. Hoffmann, H. Biologic normal values in Wistar (Jena) rats of different ages. VI. Stomach emptying time, small intestine transit time and length development of the small intestine. Z. Versuchstierkd, 1970, 12, 347-355.

694. Hoffmeister, H. E., Brunner, L., Heisig, B. & Tute, M. Treatment of postoperative secondary aldosteronism and postoperative gastroinestinal paresis with aldosterone antagonists. Med. Klin., 1970, 65, 1414-1417.

695. Hofmeister, F. & Schutz, E. Uber die automatischen Bewegungen des Magens. Arch. Exp. Pathol. Pharmakol., 1885, 20, 1-33.

696. Hogben, C. A. M. The alimentary tract. Ann. Rev. Physiol., 1960, 22, 381-406.

697. Holcomb, R. & Richards, T. W. Simultaneous records of gastric and bodily movements of puppies. Proc. Ia. Acad. Sci., 1935, 42, 176.

698. Hollander, F. & Penner, A. History and development of gastric analysis procedure. Am. J. Dig. Dis., 1939, 5, 22-25, 739-743, 786-791.

699. Holm, C. N. & Hansen, L. P. Plant fibres and duration of gastro-intestinal passage. Ugeskr. Laeger., 1975, 137, 561-565.

700. Holman, M. E. Introduction to electrophysiology of visceral smooth muscle. In C. F. Code (Ed.), Handbook of Physiology. Sect. 6, Alimentary Canal. Vol. IV. Motility. Washington, D.C.: American Physiological Society, 1968.

701. Holman, M. E. Junction potentials in smooth muscle. In E. Bulbring, A. F. Brading, A. W. Jones, & T. Tomita (Eds.), Smooth Muscle. London: Arnold, 1971.

702. Holzmann, E. M. & Sakharova, D. S. Electrical sensitivity of the human stomach (different mechanisms in production of mechanical activity). Bull. Biol. Med. Exp. U.S.S.R., 1947, 23, 116-119.

703. Homans, J. Relation of pain in gastric and duodenal ulcer to muscular activity of the stomach. Am. J. Med. Sci., 1919, 157, 74-83.

704. Hopf, H. Studien uber antagonistische Nerven. Nr. VII. Uber den hemmenden und erregenden Einfluss des Vagus auf den Magen des Frosches. Zeitschr. f. Biol., 1911, 55, 409-459.

705. Hopkins, A. The pattern of gastric emptying: A new view of old results. J. Physiol., London, 1966, 182, 144-149.

706. Horton, B. T. Pyloric measurement with special reference to pyloric block. _Am. J. Anat._, 1928, _41_, 197.

707. Horton, R. E., Ross, F. G. M. & Darling, G. H. Determination of the emptying time of the stomach by the use of entericcoated barium granules. _Br. Med. J._, 1965, _1_, 1537-1539.

708. Howard, F. A. & Sharp, D. S. The effect of intra-muscular metoclopramide on gastric emptying during labour. _Postgrad. Med. J._, 1973, _49_, Suppl. 4, 53-57.

709. Howells, T. H., Khanam, T., Kreel, L., Seymour, C., Oliver, B. & Davies, J. A. Pharmacological emptying of the stomach with metoclopramide. _Br. Med. J._, 1971, _2_, 558-560.

710. Hubel, K. A. Voluntary control of gastrointestinal function: Operant conditioning and biofeedback. _Gastroenterology_, 1974, _66_, 1085-1088.

711. Humphrey, C. S., Wilkinson, A. R. & Johnston, D. Dumping and diarrhoea after pyloroplasty and rate of gastric emptying as demonstrated by oral glucose solution. _Br. J. Surg._, 1970, _57_, 383-384.

712. Hunt, J. N. Viscosity of a test meal, its influence on gastric emptying and secretion. _Lancet_, 1954, _1_, 17-18.

713. Hunt, J. N. Influence of hydrochloric acid on gastric secretion and emptying in patients with duodenal ulcer. _Br. Med. J._, 1957, _1_, 681-684.

714. Hunt, J. N. The regulation of gastric emptying. In F. Avery Jones (Ed.), _Modern Trends in Gastro-enterology_, New York: Hoeber, 1958.

715. Hunt, J. N. Gastric emptying and secretion in man. _Physiol. Rev._, 1959, _39_, 491-533.

716. Hunt, J. N. The site of receptors slowing gastric emptying in response to starch in test meals. _J. Physiol._, London, 1960, _154_, 270-276.

717. Hunt, J. N. The osmotic control of gastric emptying. _Gastroenterology_, 1961, _41_, 49-51.

718. Hunt, J. N. Gastric emptying in relation to drug absorption. _Am. J. Dig. Dis._, 1963, _8_, 885-894.

719. Hunt, J. N. The duodenal regulation of gastric emptying. _Gastroenterology_, 1963, _45_, 149-156.

720. Hunt, J. N. & Knox, M. T. The regulation of emptying of meals containing citric acid and the salts of citric acid. J. Physiol., London, 1962, 163, 34-35.

721. Hunt, J. N. & Knox, M. T. The action of potassium oleate and potassium citrate in slowing gastric emptying. J. Physiol., London, 1964, 171, 247-253.

722. Hunt, J. N. & Knox, M. T. The effect of acids on gastric emptying. J. Physiol., London, 1965, 181, 63.

723. Hunt, J. N. & Knox, M. T. A relation between the chain length of fatty acids and the slowing of gastric emptying. J. Physiol., London, 1968, 194, 327-336.

724. Hunt, J. N. & Knox, M. T. Control of gastric emptying. Am. J. Dig. Dis., 1968, 13, 372-375.

725. Hunt, J. N. & Knox, M. T. Regulation of gastric emptying. In C. F. Code (Ed.), Handbook of Physiology. Sect. 6, Alimentary Canal. Vol. IV. Motility. Washington, D.C.: American Physiological Society, 1968.

726. Hunt, J. N. & Knox, M. T. The slowing of gastric emptying by four strong acids and three weak acids. J. Physiol., 1972, 222, 187-208.

727. Hunt, J. N. & Knox, H. T. The slowing of gastric emptying by nine acids. J. Physiol., 1969, 201, 161-179.

728. Hunt, J. N., Knox, M. T. & Oglinski, A. The effect of gravity on gastric emptying with various test meals. J. Physiol., London, 1965, 178, 92-97.

729. Hunt, J. N. & MacDonald, I. The influence of volume on gastric emptying. J. Physiol., London, 1954, 126, 459-474.

730. Hunt, J. N., MacDonald, I. & Spurrell, W. R. The gastric response to pectin meals of high osmotic pressure. J. Physiol., London, 1951, 115, 185-195.

731. Hunt, J. N. & Pathak, J. D. The osmotic effects of some simple molecules and ions on gastric emptying. J. Physiol., London, 1960, 154, 254-269.

732. Hunt, J. N. & Ramsbottom, N. Effect of gastrin II on gastric emptying and secretion during a test meal. Br. Med. J., 1967, 4, 386-387.

733. Hunt, J. N. & Spurrell, W. R. The pattern of emptying of the human stomach. J. Physiol., London, 1951, 113, 157-168.

734. Hunt, J. N. & Stubbs, D. F. The volume and energy content of meals as determinants of gastric emptying. J. Physiol., London, 1975, 245, 209-225.

735. Hunt, P. S., Connell, A. M. & Smiley, T. B. The cricopharngeal sphincter in gastric reflux. Gut, 1970, 11, 303-306.

736. Hurwitz, L., Battle, F. & Weiss, G. B. Action of the calcium antagonists cocaine and ethanol on contraction and potassium efflux of smooth muscle. J. Gen. Physiol., 1962, 46, 315-332.

737. Hutet, G. & Faucon, G. Frequence des mouvements gastr-intestinaux et systeme nerveux vegetatif. Compt. Rend. Soc. Biol., 1960, 154, 620-623.

738. Ichikawa, S. Persistalsis. J. Physiol. Soc. Jap., 1971, 33, 116-125.

739. Ichikawa, S. & Bozler, E. Monophasic and diphasic action potentials of the stomach. Am. J. Physiol., 1955, 182, 92-96.

740. Iggo, A. Tension receptors in the stomach and urinary bladder. J. Physiol., London, 1955, 128, 593-607.

741. Iggo, A. Central nervous control of gastric movements in sheep and goats. J. Physiol., London, 1956, 181, 248-256.

742. Iggo, A. Gastro-intestinal tension receptors with unmyelinated afferent fibres in the vagus of the cat. Quart. J. Exp. Physiol., 1957, 42, 130-143.

743. Iggo, A. Gastric mucosal chemoreceptors with vagal afferent fibres in the cat. Quart. J. Exp. Physiol., 1957, 42, 398-409.

744. Iluchev, D. Changes in the motility of the prepyloric region of the stomach in acute metabolic acidosis in rabbits. Folia. Med. (Plovdiv), 1968, 10, 244-249.

745. Indovina, I., Scaffidi, L. & Cataldo, M. G. Gastric motility studied with fluids labeled with TCALB 99M. Boll. Soc. Ital. Biol. Sper., 1974, 50, 487-493.

746. Ingram, P. W. & Richards, D. L. Observations on potential differences of the human stomach by a new electrode and measuring system. Gastroenterology, 1953, 25, 273-289.

747. Interone, C. V., DelFinado, J. E., Miller, B., Bombeck, C. T. & Nyhus, L. M. Parietal cell vagotomy: studies of gastric emptying and observations of protection from histamine induced ulcer. Arch. Surg., 1971, 102, 43-44.

748. Iriki, M., Tsuchiya, K., Kozawa, E., Kuramoto, T. & Miura, A. The effects of flavoxate hydrochloride on the gastrointestinal motility. Jap. J. Smooth Muscle Res., 1975, 11, 29-37.

749. Irwin, D. A. The anatomy of the Auerbach's plexus. Am. J. Anat., 1931, 49, 141-166.

750. Isenberg, J. I. & Csendes, A. Effect of octapeptide of cholecystokinin on canine pyloric pressure. Am. J. Physiol., 1972, 222, 428-431.

751. Ishikawa, M. & Ishimori, A. Symposium on gastrointestinal hormones. A clinical study of regulation of motility of digestive tract by gastrointestinal hormones. J. Jap. Soc. Intern. Med., 1974, 63, 1047-1051.

752. Itabisashi, T. & Matsumoto, H. Electrophysiological studies on the movements of the ruminant stomach. Nat. Inst. Anim. Health Quart., 1966, 6, 43-55.

753. Ito, Y., Kuriyama, H. & Sakamoto, Y. Effects of tetralthylammonium chloride on the membrane activity of guinea-pig stomach smooth muscle. J. Physiol., London, 1970, 211, 445-460.

754. Ivy, A. C. & Gray, J. S. Enterogastrone. Cold Spring Harbor Symp. Quant. Biol., 1937, 5, 405-409.

755. Iwasaki, A., Yoshida, H. & Matsushima, T. Electromyographic changes in experimental gastric ulcer. Jap. J. Smooth Muscle Res., 1974, 10, 227-229.

756. Jablonsk, A. M., Souckov, A. E., Skop, V. & Stanidis, L. Contribution to investigations of the gastric motility in patients with gastroduodenal ulcers. Sb. Lek., 1968, 70, 293-300.

757. Jablonsk, A. M., Souckov, A. E., Skop, V. & Stanidis, L. Relationship between antral motility gastric and the maximum stimulated gastric secretion in patients with gastroduodenal ulcers. Sb. Lek., 1968, 70, 301-306.

0 758. Jacobson, E. D. & Shanbour, L. L. Gastrointestinal Physiology. Baltimore: Univ. Park Press, 1974.

759. Jacoby, H. I. & Brody, D. A. Gastrointestinal action of metoclopramide. Gastroenterology, 1967, 52, 676-684.

760. Jacoby, H. I., Bass, P. & Bennett, D. R. In vivo extraluminal contractile force transducer for gastro-intestinal muscle. J. Appl. Physiol., 1963, 18, 658-665.

761. James, A. H. The Physiology of Gastric Digestion. London: Arnold, 1957.

762. James, T. G. I. Value of auscultation of acute abdomen. Practitioner, 1934, 132, 495.

763. James, W. B. & Hume, R. Action of metoclopramide on gastric emptying and small bowel transit time. Gut, 1968, 9, 203-205.

764. Janowitz, H. D. Digestive system. Ann. Rev. Physiol., 1961, 23, 153-182.

765. Jansson, G. Extrinsic nervous control of gastric motility: An experimental study in the cat. Acta Physiol. Scand., 1969, 326, 1-42.

766. Jansson, G. Vago-vagal reflex relaxation of the stomach in the cat. Acta Physiol. Scand., 1969, 75, 245-252.

767. Jansson, G. & Lisander, B. On adrenergic influence on gastric motility in chronically vagotomized cats. Acta Physiol. Scand., 1969, 76, 463-471.

768. Jansson, G. & Martinson, J. Some quantitative considerations on vagally induced relaxation of the gastric smooth muscle of the cat. Acta Physiol. Scand., 1965, 63, 351-357.

769. Jansson, G. & Martinson, J. Studies on the ganglionic site of action of sympathetic outflow to the stomach. Acta Physiol. Scand., 1966, 68, 184-192.

770. Jefferson, N. C., Arai, T., Geisel, T. & Necheles, H. Humoral factor from the brain which activates gastric motility. Science, 1964, 144, 58-59.

771. Jefferson, N. C., Dailey, U. J., Walker, L. & Nicheles, H. Phrenic nerves and gastrointestinal mechanisms. Left phrenic crush and vagotomy. Am. J. Gastroenterol., 1956, 25, 545-549.

772. Jefferson, N. C., Kuroyanagi, Y., Arai, T., Geisel, A. & Necheles, H. Extravagal gastric motor innervation. Surgery, 1965, 58, 420.

773. Jefferson, N. C., Kuroyanagi, Y., Geisel, T. & Necheles, H. Stomach contraction upon central vagus stimulation. Science, 1963, 140, 810-811.

774. Jefferson, N. C. & Necheles, H. Effect of phrenicotomy on gastrointestinal mechanisms. Proc. Soc. Exp. Biol. Med., 1948, 68, 166-169.

775. Jefferson, N. C., Walker, L. & Necheles, H. Influence of right phrenic exersis on gastric and biliary mechanisms. Am. J. Physiol., 1955, 183, 237-239.

776. Johansson, C. Studies in gastrointestinal interactions. IV. Gastric emptying of a composite meal in man. The influence of glucose. Scand. J. Gastroenterol., 1973, 8, 533-539.

777. Johansson, C. Studies of gastrointestinal interactions VII. Scand. J. Gastroenterol., 1975, 10, 33-42.

778. Johansson, C., Ekelund, K., Kulsdom, N. & Larson, I. Studies of gastrointestinal interactions. II. Calculation of gastric evacuation in an in vitro model. Scand. J. Gastroenterol., 1972, 7, 391-393.

779. Johansson, H. Gastrointestinal motility function related to thyroid activity. An experimental study in the rat. Acta Chir. Scand., 1966, Suppl, 359, 1-88.

780. Johnson, A. G. The action of metoclopramide on human gastroduodenal motility. Gut, 1971, 12, 421-426.

781. Johnson, A. G. Gastroduodenal motility and synchonization. Postgrad. Med. J., 1973, 49, Suppl. 4, 29-34.

782. Johnson, L. P., Brown, J. C. & Magee, D. F. Effect of secretin and cholecystokinin-pancreozymin extracts on gastric motility in man. Gut, 1966, 7, 52-57.

783. Johnson, L. P. & Magee, D. F. Cholecystokin-in-pancreozymin extracts and gastric motility. Surg. Gynecol. Obstet., 1965, 121, 557-562.

784. Johnson, L. P. & Magee, D. F. Inhibition of gastric motility by a commercial duodenal mucosal extract containing cholecystokinin and pancreozymin. Nature, 1965, 207, 1356-1359.

785. Johnson, L. P., Sloop, R. D., Jesseph, J. E. & Henley, E. D., Jr. Plethysmographic evidence supporting the concept of a humoral etiology of the experimental dumping syndrome. J. Surg. Res., 1962, 2, 241-244.

786. Jones, B. E. & Poulsen, J. S. Abomasal emptying rate in goats and cows measured by external counting of radioactive sodium chromate injected directly into the abomasum. Nord. Vet. Med., 1974, 26, 13-21.

787. Jones, J. H. & Morgan, D. B. Measurement of plasma-protein loss into gastrointestinal tract using 131 I-labeled proteins and oral Amberlite resin. Lancet, 1963, 1, 626-629.

788. Jones, T., Clark, J. C., Kocak, N., Cox, A. G. & Glass, H. I. Measurement of gastric emptying using the scintillation camera and 129CS. Br. J. Radiol., 1970, 43, 537-541.

789. Jordan, H. A. Voluntary intragastric feeding: Oral and gastric contributions to food intake and hunger in man. J. Comp. Physiol. Psychol., 1969, 68, 498-506.

790. Jungmann, H. & Venning, P. Radiological investigation of stomach changes following a loud auditory stimulus. Br. J. Radio., 1952, 25, 202.

791. Jungmann, H. & Venning, P. Some data on the normal variation of the stomach observed radiologically. Br. J. Radiol., 1952, 25, 25-32.

792. Jungmann, H. & Venning, P. Radiological observations of stomach changes accompanying the threat of an injection, and further data on the normal variation of peristaltic rate. Br. J. Radio., 1953, 26, 93.

793. Kabat, H., Anson, B. J., Magoun, H. W. & Ranson, S. W. Stimulation of the hypothalamus with special reference to its effect on gastrointestinal motility. Am. J. Physiol., 1935, 112, 214-226.

794. Kaiser, C., Eglin, R. E. & Zanoni, G. Gastric motility after vagotomy. X-ray cinematographic evacuation study on 6 patients with vagotomy and resection or pyloroplasty. Helv. Chir. Acta., 1972, 39, 867-874.

795. Kalbasi, H., Hudson, F. R., Herring, A., Moss, S., Glass, H. I. & Spencer, J. Gastric emptying following vagotomy and antrectomy and proximal gastric vagotomy. Gut, 1975, 509-513.

796. Kaplan, P. M. & Ossetinski, T. G. Mozochok i motorna funktsiya shlunka (The cerebellum and motor function of the stomach). Med. Exp. Kharlov, 1939, 1, 31-34.

797. Kasick, A. M. & Fein, H. D. Experimental observations on the effects of glycophrolate on the acidity of gastric secretion and on the motility of the g.i. tract. Am. J. Gastroenterol., 1963, 39, 61-68.

798. Kastle, C. Zur vergleichenden Rontgenphysiologie der Magenbewegung. Fortschr. a. d. Geb. d. Rontgenstrahlen, 1918-19, 26, 181-193.

799. Kastle, C., Rieder, H. & Rosenthal, J. The bioroentgenography of the internal organs. Arch. Roentg. Ray., 1910, 15, 3-12.

800. Katsch, G. In G. V. Bergmann and R. Staehelin (Eds.), Handbuch der inneren Medizin. Berlin: Springer, 1926.

801. Kay, A. W. & Smith, A. N. The action of atropine and hexamethonium in combination on gastric secretion and motility. Br. J. Pharmacol., 1956, 11, 231-235.

802. Kay, R. N. & Ruckenbusch, Y. Movements of the stomach compartments of a young bull during sucking. Br. J. Nutr., 1971, 26, 301-309.

803. Kay, R. N. & Ruckenbusch, Y. The effects of sucking on the motility of the forestomach in young ruminants. J. Physiol., London, 1971, 34-35.

804. Keel, H. J., Goldstein, M. L., Haubrich, W. S. & Roth, J. L. On the effect of nutritional products unsaturated fatty acids on gastric acidity, gastric motility and blood lipids. Schweiz. Med. Wochenschr., 1968, 98, 288-289.

805. Keet, A. D. The prepyloric contractions in the normal stomach. Acta Radiol., 1957, 48, 413-424.

806. Kelley, M. L., Jr., Wilbur, D. L. III, Schlegel, J. F. & Code, C. F. Deglutitive responses in the gastroesophageal sphincter of healthy human beings. J. Appl. Physiol., 1960, 15, 483-488.

807. Kelly, J. M. Metoclopramide in the vagotomized stomach. II. In dog. Postgrad. Med. J., 1973, 49, Suppl. 4, 83-87.

808. Kelling, G. Untersuchungen uber die Spannungszustande der Bauchwand, der Magen und der Darmwand. Zeitschr. f. Biol., 1903, 44, 161-258.

809. Kelly, K. A. Effect of gastrin on gastric myo-electric activity. Am. J. Dig. Dis., 1970, 15, 399-405.

810. Kelly, K. A. Gastric motility and ulcer surgery. Surg. Clin. North Am., 1971, 51, 927-934.

811. Kelly, K. A. Canine gastric motility and emptying: electric, neural and hormonal controls. In E. E. Daniel et al. (Eds.), Proc. 4th International Symposium on Gastrointestinal Motility. Vancouver: Mitchell Press, 1974.

812. Kelly, K. A. Differential responses of the canine gastric corpus and antrum to electric stimulation. Am. J. Physiol., 1974, 226, 230-234.

813. Kelly, K. A. Gastric motility after gastric operations. Surg. Annu., 1974, 6, 103-123.

814. Kelly, K. A. & Code, C. F. Effect of transthoracic vagotomy on canine gastric electrical activity. Gastroenterology, 1969, 55, 51-58.

815. Kelly, K. A. & Code, C. F. Canine gastric pacemaker. Am. J. Physiol., 1971, 220, 112-118.

816. Kelly, K. A., Code, C. F. & Elveback, L. R. Patterns of canine gastric electrical activity. Am. J. Physiol., 1969, 217, 461.

817. Kelly, K. A. & LaForce, R. C. Pacing the canine stomach with electric stimulation. Am. J. Physiol., 1972, 222, 588.

818. Kelly, K. A. & LaForce, R. C. Role of the gastric pacesetter potential defined by electrical pacing. Can. J. Physiol. Pharmacol., 1972, 50, 1017.

819. Kelly, K. A., Morely, K. D. & Wilbur, B. G. Effect of corpural and antral gastrojejunostomy on canine gastric emptying of solid spheres and liquids. Br. J. Surg., 1973, 60, 880-884.

820. Kelly, K. A., Woodward, E. R. & Code, C. F. Effect of secretin and cholecystokinin on canine gastric electrical activity. Proc. Soc. Exp. Biol. Med., 1969, 130, 1060.

821. Kelly, M. & Kennedy, T. Motility changes in the antrum after proximal gastric vagotomy. Br. J. Surg., 1975, 62, 215-220.

822. Kelly, R. E. & Rice, R. V. Ultrastructural studies on the contractile mechanism of smooth muscle. J. Cell. Biol., 1969, 42, 684-694.

823. Kemp, D. R., Herreta, R., Tsukamoto, M. & Eisenberg, M. M. Insulin-potassium effects on gastric acid secretion and antral motility in dogs. Gastroenterology, 1968, 54, 190-196.

824. Kennedy, T. Metoclopramide in the vagotomized stomach. I. In man. Postgrad. Med. J., 1973, 49, Suppl. 4, 81-82.

825. Khalikov, T. & Bulgakov, G. A. Motor evacuatory function of the gastric and small intestinal stump in patients with dumping syndromes. Sov. Med., 1970, 33, 61-63.

826. Khalikov, T. & Lokshina, K. A. The relationship between the rate of evacuation of the stomach stump, severity of the dumping syndrome and the extent of gastrojejunal anastomsis. Klin. Khir., 1969, 9, 14-17.

827. Kharchenko, N. M. & Groisman, S. D. The effect of disrupting the pyloric sphincter on gastric evacuation. Biull. Eksp. Biol. Med., 1975, 79, 25-29.

828. Kholod, A. V. & Storozhenko, I. N. Motoricity of the stomach before and after gastric resection for peptic ulcer. Sov. Med., 1971, 34, 120-125.

829. Khoromski, L. N. The motor-secretory function of the stomach in euthyroid and thyrotoxic goiter. Probl. Endokrinol. Gormonoter, 1966, 12, 27-30.

830. Khoromski, L. N. & Iurchenko, A. P. Diagnostic significance of examination of the gastric motor function during surgical interventions on the stomach and other organs of the abdominal cavity. Klin. Khir., 1966, 5, 65-68.

831. Kihara, T., Isihara, Y. & Ota, A. The effect of a prifinim bromide (1, 1-diethyl-3-diphenyl methylene-2-methylpyrrolidium bromide) on the gastric antral motility in man. Arzneim. Forsch., 1970, 20, 762-767.

832. Kilby, J. O. & Griffith, C. A. Selective and total vagotomy with suprapyloric antrectomy: A comparative study of gastric secretion and emptying in dogs. Surgery, 1971, 69, 702-707.

833. Kilby, J. O. & Griffith, C. A. The relationships of gastric transection and vagotomy to gastric emptying. Surgery, 1971, 69, 633-636.

834. Killen, D. A. & Symbas, P. N. Effect of preservation of the pyloric sphincter during antrectomy on postoperative gastric emptying. Am. J. Surg., 1962, 104, 836-842.

835. Killian, J. A. & Marsh, M. E. The effect of fat on the gastric responses to foods. Oil & Soap, 1945, 22, 250.

836. Kimura, K. Photo-optical observation on motility of residual stomach after proximal gastrectomy. Jap. J. Smooth Muscle Res., 1972, 8, 112-125.

837. Kimura, M., Dobeta, H. & Ono, Y. Physiological study on autogenic training. J. of Jap. Psychosom. Soc., 1964, 4, 208-211.

838. King, J. E. Postoperative abdominal auscultation. Am. J. Obstet. and Gynec., 1930, 19, 237.

839. King, J. E. Postoperative abdominal auscultation. N.Y. State J. Med., 1931, 31, 80.

840. King, J. L. & Connet, H. Contributions to the physiology of the stomach. XXIX. The gastric hunger contractions of the normal and decerebrate guinea-pig. Am. J. Physiol., 1915, 39, 123-130.

841. Kipiani, T. I. Classification of the motor-evacuation function of the stomach in experiments with mineral waters. Vopr. Kurortol. Fizioter. Lech. Fiz. Kult., 1966, 31, 366-368.

842. Kirov, A. A., Popov, G. V., Poliakov, D. I., Popov, I. G. & Vorobev, I. A. Motor function of the gastrointestinal tract in the early period following a draining operation on the stomach in combination with truncal and selective vagotomy (an experimental study). Klin. Khir., 1974, 86-88.

843. Kirschner, M. & Mangold, E. Die motorische Funktion des Sphincter pylori und des Antrum pylori beim Hunde nach der queren Durchtrennung des Magens. Mitt. a. d. Grenzgeb. d. Med. u. Chir., 1911, 23, 446-494.

844. Kirwan, W. O. & Smith, A. N. Gastrointestinal transit estimated by an idotope capsule. Scand. J. Gastroenterol., 1974, 9, 763-766.

845. Kit, O. N. Motor-secretory function of the resected stomach in patients with peptic ulcer. Vestn. Khir., 1973, 11, 8-10.

846. Klein, E. Gastric motility. I. The origin and character of gastric peristalsis. Arch. Surg., 1926, 12, 571-582.

847. Klimov, P. K. Synchronization and time sequence of the motility of the stomach, duodenum, biliary pathways and pancreas. Cesk. Gastroenterol. Vyz., 1975, 29, 176-183.

848. Klimov, P. K., Nozdrachev, A. D. & Chernigovskii, V. N. Afference in the sympathetic pathways of the stomach related to its motor-evacuatory activity. Fiziol. Zh. S.S.S.R., 1973, 59, 1268-1278.

849. Klimov, P. K., Slobozhandin, A. D., Fokina, A. A. & Iutin, A. I. Humoral regulation of the gastrointestinal motor function. Fiziol. Zh. S.S.S.R., 1971, 57, 270-275.

850. Koch, C. R. & Stunkard, A. J. Obesity, age and gastric hunger contractions. J. Appl. Physiol., 1960, 15, 133-134.

851. Kochemasova, N. G., Shovba, M. F. & Boev, K. Electrical and contractile activities of the stomach muscles from the stomach in cats. Dokl. Bolg. Akad. Nauk., 1969, 22, 1437-1440.

852. Kock, N. G. An experimental analysis of mechanisms engaged in reflex inhibition of intestinal motility. Acta Physiol. Scand., 1959, 47, 5-54.

853. Kohatsu, S. A study of the human electrogastrogram using cutaneous electrodes. Jap. J. Smooth Muscle Res., 1968, 4, 148-150.

The cutaneous electrodes, despite certain disadvantages, are suitable electrodes to study human subjects. Electrogastrogram, adequate for analysis of rate changes, can be obtained. The advantages are: there is no discomfort to the patient; long term studies can be performed in comfort; there is no stimulation of the mucosa. Mean fasting rate of the normal human stomach is 2.9 with a standard deviation of .3 cycles per minute. The feeding response of the human stomach has a definite pattern. Following an initial drop in rate almost immediately after beginning ingestion of food, there is a rise within 28 minutes above the fasting rate with a gradual decrease towards the fasting rate.

854. Kohatsu, S. Human electrogastrography. Jap. J. Smooth Muscle Res., 1970, 6, 129-132.

The cutaneous electrodes are simple to use and there is no discomfort for the subject. Long-term studies are possible with these electrodes. The quality of the recordings is not as good as those obtained from implanted or swallowed electrodes, but this study demonstrates that the silver-silver chloride cutaneous electrodes produce adequate, accurate electrogastrograms for rate analyses. In addition, eating produces a characteristic response of the gastric rate, and rate analysis is important in human electrogastrography.

855. Kohatsu, S. The current status of electrogastrography. Klin. Wochenschr., 1970, 48, 1315-1319.

Most of our present concepts and basic knowledge of gastric muscle electrophysiology are based on animal studies. A brief review of those studies which are relevant in understanding the human studies are presented. The studies of human electrogastrography and the present concepts of clinical electrogastrography are reviewed. The various types of electrodes used to record the electrogastrograms, and their advantages and disadvantages are discussed. The physiological data presently available from electrogastrograms are presented.

856. Kolling, K. Motor activity of the rumino-reticulum. I. Character of movements of different parts and their time relations. Zentralbl. Veterinaermed., 1974, [A] 21, 132-141.

857. Kolling, K. Motor activity of the rumino-reticulum. 2. Movements of different sections and their significance for regurgitation. Zentralbl. Veteringermed., 1974, [A] 21, 268-278.

858. Kolossow, N. G. & Sapojnikowa, D. L. Observations concernant l'innervation de la voie digestive chez les ruminants. Trab. Lab. Invest. Biol. Univ. Madrid, 1933, 28, 345-368.

859. Komarchev, K. G., Firsova, P. P. & Pechatnikova, E. A. Motor-evacuatory function of the stomach following Billroth II surgery in our modification. Vestn. Khir., 1971, 107, 24-27.

860. Konegawa, R., Kuramoto, M. & Watanabe, H. The role of the pyloric branch of the vagus nerve during astric motions. Jap. J. Smooth Muscle Res., 1974, 10, 230-231.

861. Korkuschko, O. W., Jakimenko, D. M., Wolostschenko, I. I. & Kotko, D. N. Age-induced changes of the motor and secretory functions of the stomach as initial criteria in the diganosis of pathologic conditions. Z. Alternasforsch., 1972, 25, 263-269.

862. Korotko, G. F. & Abliazov, A. A. Regulation of gastric evacuatory activity. Fiziol. Zh. S.S.S.R., 1973, 59, 337-343.

863. Kosaka, T. & Lim, R. K. S. On the mechanism of inhibition of gastric motility by fat. An inhibitory agent from the intestinal mucosa. Chin. J. Physiol., 1933, 7, 5-12.

864. Koster, N. & Madsen, P. The intragastric pressure before and immediately after truncal vagotomy. Scand. J. Gastroenterol., 1970, 5, 381-383.

865. Kosterlitz, H. W. Intrinsic and extrinsic nervous control of motility of the stomach and the intestines. In C. F. Code (Ed.), Handbook of Physiology. Sect. 6, Alimentary Canal. Vol. IV. Motility. Washington, D.C.: American Physiological Society, 1968.

866. Kottegoda, S. R. An analysis of possible nervous mechanisms involved in the peristaltic reflex. J. Physiol., London, 1969, 200, 687-712.

867. Koval, V. P. & Diagai, I. V. The effect of carbohydrate loading on the motor function of the gastric stump. Khirurgiia, Mosk., 1971, 47, 64-68.

868. Kowalewski, K. Effect of calcium infusion on gastric function in rats. Proc. Soc. Exp. Biol. Med., 1968, 127, 1134-1137.

869. Kowalewski, K. & Zajac, S. Effect of hemorrahage on myoelectric activity of stomach and duodenum in dogs. Surg. Gynecol. Obstet., 1975, 141, 227-231.

870. Kowalewski, K. & Zajac, S. Electrical and mechanical activity of the isolated canine stomach perfused with homologous in vitro oxygenated blood. Pharmacology, 1975, 13, 448-457.

871. Kowalewski, K., Zajac, S. & Kolodej, A. Effect of ischemic anoxia on myoelectrical and mechanical activity of the totally isolated canine stomach. Surg. Gynecol. Obstet., 1975, 140, 187-192.

872. Kowalewski, K., Zajac, S. & Kolodej, A. Effect of release of endogenous gastrin on myoelectrical and mechanical activity of isolated canine stomach. Pharmacology, 1975, 13, 56-64.

873. Kowalewski, K., Zajac, S. & Kologsdej, A. The effect of drugs on the electrical and mechanical activity of the isolated porcine stomach. Pharmacology, 1975, 13, 86-95.

874. Kowalski, R. Role of emotional factors in digestive tract motility in children. Ann. Radiol., (Paris), 1968, 11, 545-550.

875. Kradinov, A. I. Motor-evacuatory function of stomach after resection with jejunogastroplasty. Vrach. Delo., 1967, 2, 1-4.

876. Kradinov, A. I. Roentgenofunctional state of the stomach in the late periods after resection with jejunogastroplasty. Vestn. Rentgenol. Radiol., 1970, 45, 7-13.

877. Krapivin, B. V. & Chernin, V. V. Study of the motor function of the stomach by the electrogastrographic method in patients with peptic ulcer. Sov. Med., 1967, 30, 57-60.

In an acute period of exacerbation of an ulcer a tendency towards an increase of peristaltic activity of the stomach is observed. The appearance/phenomenon of irregular biopotentials and an increase in their frequency/mi is noted. During a subacute period of exacerbation of the illness in the majority of cases the peristaltic activity of the stomach is lowered, a hypokinetic type of EGG appeares. The irregularity of the appearance of biopotentials and their frequency/mi increases. In a period of healing of the ulcer and of the liquidation/ elimination of the symptoms of exacerbation, the motor function of the stomach normalizes significantly. A complex/complicated ulcer proceeds on a background of a pronounced decrease of the motor activity of the stomach. Fifteen days after surgery the peristaltic activity of the stomach essentially is distinguished/ differs from before the operation.

878. Krasil'nikov, L. G. Linicheskope znochenize elektrogastrogrefil. (Clinical importance of electrogastrography). Sov. Med., 1960, 24, 107-114.

M. A. Sobakin developed a new method for the sensing of bioelectric currents from the anterior abdominal wall with the utilization of a new electrogastrograph--EGS-1 and EGS-2--created by him together with the engineers M. A. Gurevich, S. M. Iakovlev, and L. I. Mishkin. In the methodology proposed by the author, the experience and the failures of both native and foreign electrophysiologists who studied electrogastrography are taken into account. The apparatus records bio-electric currents from the point of projection of the stomach on the anterior abdominal wall, and has a high, selective sensitivity; its frequency range is from 0.04 to 0.02 hertz, and for the limits of the service band the frequency-response curve has steep slopes to the side of high and low frequencies; this permits the registration and study, in a rather pure form, of bio-electric currents, corresponding to the peristaltic movements of the stomach, which originate in the period of digestion with the use of any (in qualitative and quantitative measure) experimental meal. The new methodology and apparatus allows one to hope that the lack of correlation in the clinic between the study of the secretory function of the stomach in the period of digestion, and the motor function in the empty stomach, will be eliminated.

879. Krasil'nikov, L. G. & Fishzon-ryss, I. Comparative study of mechano-electro-graphic recordings of peristalsis in human stomach. Novost. Med. Tekhn, 1962, 1, 66-71.

In order to understand these peculiarities, it is essential to note that the avacuation of fluid from the stomach is controlled chiefly by the tone and motor activity of its pylorus-antrum section. Active peristalsis of the core/center of the stomach in this period is usually not observed (19). Consequently, the projections of the EGG in the given cases can be attributed to the current of activity which is brought about by the termination of the existing part of the stomach. With the termination of the evacuation of the basic mass of fluid from the stomach, the projections of the EGG lose height significantly (Figure 4b). Summarizing all the foregoing, we will notice that the data obtained allows us to advance a judgement concerning the genesis of the so-called "spontaneous vascillations" of the electropotential of the stomach. The simultaneous recording of the EGG and WGG including the conditions which exclude the influence of the acidic factor allows us to reveal the correspondence between the electrical and mechanical indices of peristalsis of the pylorus-antrum section of the stomach. The results of pharmacological investigation also confirm the dependence of the projections of the EGG on the motor activity of the stomach. What is stated here does not contradict the facts which were obtained in a comparison of the electrographic method of M. A. Sobakin with the mechanogastrographic recording obtained with the method of Bikov-Kyrtsin. For this reason it appears to us that the time has come to deny the empty and indeterminate term "spontanious vascillations of the biopotential of the stomach" and to call the vascillations "peristaltic projections of the electrogastrogram". This will bring about the necessary definiteness in the interpretation of the EGG and will promote mutual understanding of physiologists and clinicians on this issue.

880. Krasil'nikov, L. G. Interpretation of electrogastrogram and its variants in healthy persons during digestion. Fed. Proc., (Trans. Suppl.), 1964, 23, 901-904.

881. Krasnov, E. A. Rhythmical fluctuations in the biopotential of the neuromuscular apparatus of the stomach and small intestine. Vopr. Eksperim. Klin. Khir., 1961, 7, 118-136.

882. Krasnykh, I. G. & Tiutin, L. A. Motor-evacuatory function of the gastrointestinal tract in dogs in conditions of prolonged hypokinesia. *Kosm. Biol. Med.*, 1973, *7*, 40-45.

883. Krejs, G. J. & Blum, A. L. Proceedings: Gastric emptying of organic acids in the dog. *Acta Hepatogastroenterol.*, (Stuttg), 1975, *22*, 63.

884. Krishnamoorthy, M. S. & Nayar, C. K. A comparative study on the effects of pilocarpine and physostigmine on gastrointestinal motility in dogs. *Indian Vet. J.*, 1967, *44*, 1044-1051.

885. Kramer, P. Antimotility activity of equitoxic doses of anticholinergic agents. *Am. J. Dig. Dis.*, 1960, *5*, 1019-1029.

886. Kronborg, O. & Madsen, P. Gastric emptying rate and acid secretion after truncal vagotomy and pyloroplasty for duodenal ulceration. *Scand. J. Gastroenterol.*, 1972, *7*, 515-518.

887. Krzywanek, F. W. Vergleichende intersuchungen uber die mechanik der vandauung. I. Untersuchgen am hund. (Comparative studies of the mechanics of digestion. I. Studies on the dog). *Pfluger's Arch. f. d. gesammt. Physiol.*, 1927, *215*, 481-518.

888. Kuiper, D. H. Gastric bezoar in a patient with myotonic dystrophy. A review of the gastrointestinal complications of myotonic dystrophy. *Am. J. Dig. Dis.*, 1971, *16*, 529-534.

889. Kulakov, P. E. Comparative evaluation of the motor-excretory and secretory function of the stomach after jejunoplasty and Billroth II operation. *Khirurgiia*, (Mosk.), 1965, *41*, 92-98.

890. Kuo, W. C. & Wu, K. H. Early postoperative gastrointestinal motility. *Chin. Med.*, 1974, *7*, 442-445.

891. Kure, K., Ichiko, K. I. & Ishikawa, J. On the spinal parasympathetic. Physiological significance of the spinal parasympathetic system in relation to the digestive tract. *Quart. J. Exp. Physiol.*, 1931, *21*, 1-19.

892. Kuriyama, H., Osa, T. & Tassaki, H. Electrophysiological studies of the antrum muscle fibers of the guinea pig stomach. *J. Gen. Physiol.*, 1970, *55*, 48-62.

893. Kuriyama, H. & Tomita, T. Effect of polarizing currents on the membrane activity of single smooth muscle cells. J. Physiol., London, 1964, 173, 10-11.

894. Kuroyanagi, Y., Jefferson, N. C., Geisel, T. & Necheles, H. Chronic vagus stimulation of the stomach. Surgery, 1963, 53, 348-351.

895. Kurygin, A. A. & Groisman, S. D. Treatment of diarrhea and stomach motor disorders developing after vagotomy. Klin. Med., (Mosk.), 1971, 49, 65-68.

896. Kuzin, M. I., Shirokova, K. I., Chistova, S. V. & Smakov, G. M. Periodic motor activity and secretory function of the empty stomach in myasthenia. Klin. Med., (Mosk.), 1970, 48, 130-138.

897. Kuznetsov, N. N. & Belitskii, A. E. Functional state of the gastric stump after gastrectomy for peptic ulcer. Sov. Med., 1970, 33, 64-67.

898. Kuznetsov, N. N. & Zheleznyi, V. I. Comparative assessment of the motor-evacuatory function of the gastric stump after resection of various portions of the stomach. Klin. Khir., 1968, 1, 73-74.

899. Kwong, N. K., Brown, B. H., Whittaker, G. E. & Duthie, H. L. Electrical activity, motility, and acid secretion of the human stomach. Br. J. Surg., 1970, 57, 384.

900. Kwong, N. K., Brown, B. H., Whittaker, G. E. & Duthie, H. L. Electrical activity of the gastric antrum in man. Br. J. Surg., 1970, 57, 913-916.

901. Kwong, N. K., Brown, B. H., Whittaker, G. E. & Duthie, H. L. Response of the electrical activity, motor activity, and acid secretion of the human stomach to pentagastrin and histamine stimulation. Scand. J. Gastroenterol., 1971, 6, 145-153.

902. Kwong, N. K., Brown, B. H., Whittaker, G. E. & Duthie, H. L. Effects of gastrin I, secretin, and chlecystokinin-pancreozymin on the electrical activity, motor activity, and acid output of the stomach in man. Scand. J. Gastroenterol., 1972, 7, 161.

903. Lab, A. H., Lanfranchi, G. A., Barbara, L., Trombini, G., Bartolotti, M. & Miglioli, M. Effects of a test of psycho-visceral reactivity on gastro-intestinal motility. Arch. Fr. Mal. App. Dig., 1973, 62, 473-480.

904. La Barre, J. & Destree, P. L'influence des variations glycemique sur la motilite gastrique. Arch. intern. de physiol., 1931, 33, 243-256.

905. Lagemann, K. & Hoffmann, K. Gastric motility during rest and physical stress. Fortschr. Geb. Roentgenstr. Nuklearmed., 1971, 115, 817-821.

906. Lagerlof, H. O., Rudewald, M. B. & Perman, G. The neutralisation process in the duodenum and its influence on gastric emptying in man. Acta Med. Scand., 1960, 168, 269-284.

907. Lalich, J., Harrin, R. C. & Meek, W. J. Effect of a previous distention of the intestine on reflex inhibition of gastric motility. Proc. Soc. Exp. Biol. Med., 1936, 34, 29-31.

908. Lalich, J., Meek, W. J. & Herrin, R. C. Reflex pathways concerned in inhibition of hunger contractions by intestinal distention. Am. J. Physiol., 1936, 115, 410-414.

909. Lalich, J., Youmans, W. B. & Meek, W. J. Insulin and gastric motility. Am. J. Physiol., 1937, 120, 554-558.

910. Lane, B. P. & Rhodin, J. A. G. Cellular interrelationship and electrical activity in two types of smooth muscle. J. Ultrastruct. Res., 1964, 10, 470-488.

911. Lanfranchi, G. A., Massetti, L., Bortolotti, M., Barbara, L., Miglioli, M. & Sarti, A. Effect of chemical mediators of nervous regulation on the gastro-duodenal electric activity in dogs. Arch. Ital. Mal. Appar. Dig., 1969, 36, 144-165.

912. Langer, L. Gastric evacuation after vagotomy during the first days after surgery. Rozhl. Chir., 1972, 51, 529-534.

913. Langley, J. N. On inhibitory fibers in the vagus for the end of the aesophagus and the stomach. J. Physiol., 1898, 23, 407-414.

914. Laplace, J. P. Omaso-abomasal motility and feeding behavior in sheep: A new concept. Physiol. Behav., 1970, 5, 61-65.

915. Lauber, H. Znr Klinik der Bewegnngsrhythmik des Magens. Ztschr. f. d. ges. exp. Med., 1930, 74, 586-598.

916. Laughton, N. B. The effects of the stomach of stimulation of the dorsal vagus nuclei. Am. J. Physiol., 1929, 89, 18-23.

917. Lawson-Smith, C., Ralphs, D. N., Bloom, S. R. & Hobsley, M. Proceedings: Duodenal ulceration, gastric emptying, enteroglucagon release and abnormal oral glucose tolerance. Br. J. Surg., 1975, 62, 154.

918. Lebensohn, J. E. Oculovisceral reflexes. An oculogastric reflex experimentally demonstrated. Am. J. Ophth., 1929, 12, 562-568.

919. Leek, B. F. & van Miert, A. S. An analysis of the pyrogen-induced inhibition of gastric motility in sheep. J. Physiol., London, 1971, 215, 28-29.

920. Leia, I. & Danilans, A. Acid formation and evacuation from the stomach in ulcers of the duodenum. Vestn. Khir., 1974, 112, 46-49.

921. Lenz, H., Hemmati, A. & Ghavami, P. Physiology and physiopathology of the emptying of the stomach. Ann. Radiol., Paris, 1967, 10, 689-697.

922. Lewis, R. A. & Said, D. Influence of posture on gastric function. Arch. Int. Pharmacodyn., 1963, 141, 577-583.

923. Liberman, S. S. & Sharova, S. A. A comparison of the effect of the tricyclic antidepressants azaphen and imizin on the gastrointestinal tracts of experimental animals. Farmakol. Toksikol., 1975, 38, 29-32.

924. Lilie, D. The electrogastrogram. In R. F. Thompson & M. M. Patterson (Eds.), Bioelectric Recording Techniques. New York: Academic Press, 1974.

925. Lim, R. K. S. Observations on the mechanism of the inhibition of gastric function by fat. Quart. J. Exp. Physiol., 1933, 23, 263-268.

926. Lind, J. F., Duthie, H. L., Schlegel, J. F. & Code, C. F. Motility of the gastric fundus. Am. J. Physiol., 1961, 201, 197-202.

927. Linkens, D. A. & Cannell, A. E. Interactive graphics analysis of gastrointestinal electrical signals. IEEE Trans. Biomed. Eng., 1974, 21, 335-339.

928. Lipshutz, W. & Cohen, S. Interaction of gastrin I and secretin on gastrointestinal circular muscle. Am. J. Physiol., 1972, 222, 775.

929. Lisander, B. The hypothalamus and vagally mediated gastric relaxation. Acta Physiol. Scand., 1975, 93, 1-9.

930. Lisander, B. & Martner, J. Influences on gastrointestinal and bladder motility by the fastigial nucleus. Acta Physiol. Scand., 1974, 90, 792-794.

931. Lish, P. M. Some pharmacologic effects of dioctyl sodium sulfo succinate on the gastrointestinal tract of the rat. Gastroenterology, 1961, 41, 580-584.

932. Lish, P. M., Clark, B. B. & Robbins, S. I. Effects of some physiologic substances on gastrointestinal propulsion in the rat. Am. J. Physiol., 1959, 197, 22-26.

933. Lish, P. M. & Peters, E. L. Antagonism of insulin-induced gastrointestinal hypermotility in the rat. Proc. Soc. Exp. Biol. Med., 1957, 94, 664-668.

934. Littman, A., Fox, B. W., Schoolman, H. M. & Ivy, A. C. Lethal effect of intragastric irradiation in dog. Am. J. Physiol., 1953, 174, 347-351.

935. Loew, E. R. & Patterson, T. L. The reflex influence of the lower portion of the large gut on tonus and movements of the empty stomach in dogs. Am. J. Physiol., 1935, 113, 89-90.

936. Longimo, F. H., Grimson, K. S., Chittum, J. R. & Metcalf, B. H. Orally effective quarternary anime, pathine capable of reducing gastric motility and secretions. Gastroenterology, 1950, 14, 301-313.

937. Longo, A. Influenza della fatica fisica e mentale sulla funzionalita gastrica. Ass. Di. Med. Appl., 1932, 3, 210-214.

938. Lorber, S. H., Komarov, S. A. & Shay, H. Effect of sham-feeding on gastric motor activity in the dog. Fed. Proc., 1949, 8, 99.

939. Lorber, S. H., Komarov, S. A. & Shay, H. Effect of sham feeding on gastric motility of the dog. Am. J. Physiol., 1950, 162, 441-451.

940. Lorber, S. H. & Machella, T. E. Effect of syntropan on motor activities of the human gastrointestinal tract and on gastric acidity. Gastroenterology, 1949, 12, 57-69.

941. Lorber, S. H. & Shay, H. Technical and physiological considerations in measuring gastrointestinal pressures in man. Gastroenterology, 1954, 27, 478.

942. Lorber, S. H. & Shay, H. The duodenal mechanism in the control of gastric motility. Gastroenterology, 1956, 31, 117-130.

943. Lorber, S. H. & Shay, H. Effects of insulin and glucose on gastric motor activity of dogs. Gastroenterology, 1962, 43, 564-574.

944. Lortat-Jacob, J. L., Cornet, A., Rene, L., Guerre, J. & Blanc, L. Dyskinesia of the stomach and small intestine in vagotomized patients. Bull. Mem. Soc. Med. Hop., Paris, 1967, 118, 187-193.

945. Louckes, H., Brody, D. A. & Quigley, J. P. Action of the pyloric sphincter as recorded by the pyloric inductograph. Fed. Proc., 1949, 8, 100.

946. Louckes, H. & Quigley, J. P. Effect of procaine hydrochloride on the pyloric sphincter. Am. J. Physiol., 1960, 198, 1329-1332.

947. Louckes, H., Quigley, J. P. & Kersey, J. Inductograph method for recording muscle activity, especially pyloric sphincter physiology. Am. J. Physiol., 1960, 199, 301-310.

948. Louckes, H., Quigley, J. P. & Romans, W. E. Registration of sound produced by passage of chyme through pyloric sphincter; Relation to process of gastric evacuation & pyloric sphincter contraction. Am. J. Physiol., 1952, 171, 745.

949. Louckes, H., Quigley, J. P. & Romans, W. E. Sounds produced at the pyloric sphincter. Fed. Proc., 1953, 12, 91.

950. Luckhardt, A. B. & Carlson, A. J. Contributions to the physiology of the empty stomach. XVII. On the chemical control of the gastric hunger mechanism. Am. J. Physiol., 1914, 36, 37-46.

951. Luckhardt, A. B. Contributions to the physiology of the empty stomach. XXXII. The effect of dreaming on the gastric hunger contractions. Am. J. Physiol., 1916, 39, 330-334.

952. Luckhardt, A. B., Phillips, H. T. & Carlson, A. J. Contributions to the physiology of the stomach. II. The control of the pylorus. Am. J. Physiol., 1919, 50, 57-66.

953. Ludwick, J. R., Wiley, J. N. & Bass, P. Pyloroplasty and vagotomy; early effects on antral and duodenal contractile activity. Arch. Surg., 1969, 99, 533-559.

954. Ludwick, J. R., Wiley, J. N. & Bass, P. Gastric emptying following Finney pyloroplasty and vagotomy. Am. J. Dig. Dis., 1970, 15, 347-352.

955. Lui, J., Prosser, C. L. & Job, D. D. Ionic requirements of slow waves and spikes. Am. J. Physiol., 1969, 217, 1542.

956. Lukash, W. M. Fractional gastric analysis. A study of secretory and motor function of the stomach. Am. J. Gastroenterol., 1968, 49, 318-327.

957. Lulu, D. J., Lawson, L. J. & Dragstedt, L. R. Heineke-Mikulicz and Finney pyloroplasties. Effect on gastric emptying before and after vagotomy. Arch. Surg., 1971, 102, 512-515.

958. Lyons, W. S., Ellis, F. H., Jr. & Olsen, A. M. Gastroesophageal "sphincter" mechanism: Review. Proc. Staff Meetings Mayo Clinic, 1956, 31, 605-614.

959. Lysak, Z. A. Bioelectrical activity of the stomach stump after stomach resection. Khirurgiia, (Mosk.), 1974, 4, 62-97.

960. Lytkin, M. I. & Kataeva, G. A. The motor function of various organs of the gastrointestinal tract following selective vagotomy and pyloroplasty in duodenal ulcer patients. Vestn. Khir., 1972, 108, 13-16.

961. McCance, R. A., Prior, K. M. & Widdowson, E. M. Radiological study of rate of brown and white bread through digestive tract of man. Br. J. Nutr., 1953, 7, 98-104.

962. McCoy, E. J. & Bass, P. Chronic electrical activity of gastroduodenal area: Effects of food and certain catecholamines. Am. J. Physiol., 1963, 205, 539-545.

963. McCrea, E. D. & MacDonald, A. D. The action of drugs upon the movements of the stomach. Quart. J. Exp. Physiol., 1928, 19, 161-170.

964. McCrea, E. D. & McSwiney, B. A. The effect of stimulation of the vagi on the pyloric region of the stomach. J. Physiol., London, 1926, 61, 28-34.

965. McCrea, E. D. & McSwiney, B. A. The effect on the stomach of stimulation of the peripheral end of the splanchnic nerve. Quart. J. Exp. Physiol., 1928, 18, 301-313.

966. McCrea, E. D., McSwiney, B. A., Morison, J. W. & Stopford, J. S. B. The normal movements of the stomach. Quart. J. Exp. Physiol., 1924, 14, 379-397.

967. McCrea, E. D., McSwiney, B. A. & Stopford, J. S. B. The effect on the stomach of stimulation of the peripheral end of the vagus nerve. Quart. J. Exp. Physiol., 1925, 15, 201-233.

968. McCrea, E. D., McSwiney, B. A. & Stopford, J. S. B. Effects of section of the vagi nerves on the motor activity of the stomach. J. Physiol., 1925, 60, Proc., 29.

969. McCrea, E. D., McSwiney, B. A. & Stopford, J. S. B. Effects of the section of the vagi nerves on the motor activity of the stomach. Quart. J. Exp. Physiol., 1926, 16, 195-203.

970. McDonough, F. E. & Schneider, M. Effect of motion on roentgenographic appearance of stomach and small bowel. Gastroenterology, 1944, 2, 32-45.

971. McDougal, M. D. & West, G. B. The inhibition of the peristaltic reflex by sympathomimetic amines. Br. J. Pharmacol., 1954, 9, 131-137.

972. McGarry, J. The anti-emetic effect of metoclopramide with an account of some experiments to assess the effect of the drug on gastric emptying in early labour. Postgrad. Med. J., 1973, 49, Suppl. 4, 51-52.

973. McGill, C. The structure of smooth muscle in the resting and in the contracted condition. Am. J. Anat., 1909, 9, 493-545.

974. McGown, E. L. & Suttie, J. W. Influence of fat and fluoride on gastric emptying of rats. J. Nutr., 1974, 104, 909-915.

975. McIntyre, J. A., Dietel, M., Baida, M. & Jalil, S. The human electrogastrogram at operation: A preliminary report. Can. J. Surg., 1969, 12, 275-284.

976. McKelvey, S. T. Gastric incontinence and post-vagotomy diarrhoea. Br. J. Surg., 1970, 57, 741-747.

977. McKelvey, S. T., Connell, A. M. & Kennedy, T. L. Gastric emptying and transit time as factors in postvagotomy diarrhea. Gut, 1969, 10, 1047.

978. McSwiney, B. A. Innervation of the stomach. Physiol. Rev., 1931, 11, 479-514.

979. McSwiney, B. A. & Brown, G. L. Reversal of the action of adrenaline. J. Physiol., 1926, 62, 52-64.

980. McSwiney, B. A. & Robson, J. M. The sympathetic innervation of the stomach, II. The effect of stimulation of the periarterial nerves on the stomach and small intestine. J. Physiol., London, 1931, 71, 194-200.

981. McSwiney, B. A. & Robson, J. M. The sympathetic innervation of the stomach. III. The interaction of the vagus and sympathetic nerves. J. Physiol., 1931, 73, 141-150.

982. McSwiney, B. A. & Robson, J. M. The sympathetic innervation of the stomach, III. The interaction of the vagus and sympathetic nerves. J. Physiol., London, 1931, 73, 141-150.

983. McSwiney, B. A. & Spurrell, W. R. The influence of osmotic pressure on the emptying time of the stomach. J. Physiol., London, 1933, 79, 437-442.

984. McSwiney, B. A. & Spurrell, W. R. The outline method for investigating gastric motility. J. Physiol., London, 1933, 79, 423-436.

985. MacDonald, I. Gastric activity during the menstrual cycle. Gastroenterology, 1956, 30, 602-607.

986. MacDonald, R. M., Ingelfinger, F. J. & Belding, H. W. Late effects of total gastrectomy in man. New Engl. J. Med., 1947, 237, 887.

987. Machella, T. E., Hodges, H. H. & Lorber, S. H. Restoration of gastric motility by urethane of B-methyl choline after section of vagus nerves for peptic ulcer. Gastroenterology, 1947, 8, 36-51.

988. Machella, T. E. & Lorber, S. H. Gastrointestinal motility following vagotomy and use of urechochline for control of certain undesirable phenomena. Gastroenterology, 1948, 11, 426-441.

989. Mack, G., Allen, L. S. & Morton, H. S. The electrogastrograph and the effect of drugs. Can. Serv. Med. J., 1955, 11, 17-24.

990. MacLachlan, P. L. Effect of anoxic anexia on gastric emptying time of rats fed corn oil. Proc. Soc. Exp. Biol. Med., 1946, 63, 147-148.

991. MacLeod, I. B. The effects of vagal section. J. Royal Coll. Surg. Edinburgh, 1974, 19, 257-262.

992. Madsen, P., Kronborg, O. & Feldt-Rasmussen, K. The gastric emptying and small intestinal transit after highly selective vagotomy without drainage and selective vagotomy without drainage and selective vagotomy with pyloroplasty. Scand. J. Gastroenterol., 1973, 8, 541-543.

993. Magaribuchi, T., Ito, Y. & Kuriyama, H. Effects of rapid cooling on the mechanical and electrical activities of smooth muscles of guinea-pig stomach and talnia coli. J. Gen. Physiol., 1973, 61, 323-341.

994. Magee, D. F. Gastrointestinal Physiology. Springfield, Ill.: Charles C. Thomas, 1962.

995. Maisky, V. B. Comparative assessment of peristalsis in the diagnosis of carcinoma of the stomach. Vestn. Rentgenol. Radiol., 1967, 42, 42-49.

996. Maki, T. & Nelsen, T. S. Gastrointestinal electromyography. Science, 1969, 164, 334.

997. Malholta, S. L. Effect of saliva on gastric emptying. Am. J. Physiol., 1967, 213, 169-173.

998. Mann, C. V., Code, C. F. & Ellis, F. H., Jr. Intrinsic mechanisms controlling the mammalian gastroesophageal sphincter. Rendic. R. Gastroenterol., 1969, 1, 126.

999. Mann, C. V., Code, C. F., Schlegel, J. F. & Ellis, F. H., Jr. Intrinsic mechanisms controlling the mammalian gastro-esophageal sphincter deprived of extrinsic nerve supply. Thorax, 1968, 23, 634-639.

1000. Manville, I. A. & Chuinard, E. G. Studies on the gastric hunger mechanism. Am. J. Dig. Dis. Nutr., 1934, 1, 688-693.

1001. Margolin, S. G. The behavior of the stomach during psychoanalysis: A contribution to a method of verifying psychoanalytic data. Psychoanal. Quart., 1951, 20, 349-373.

1002. Margolin, S. G. Centrally mediated inhibition of gastrointestinal propulsive motility by morphine over a non-neural pathway. Proc. Soc. Exp. Biol. Med., 1963, 112, 311-315.

1003. Marik, F. & Code, C. F. Control of the interdigestive myoelectric activity in dogs by the vagus nerves and pentagastrin. Gastroenterology, 1975, 69, 387-395.

1004. Marsboom, R. & van Ravestyn, C. Effect of benzetimide (R 4929) on ruminal and intestinal motility in sheep. Br. Vet. J., 1971, 127, 264-270.

1005. Marshall, F. N., Pittinger, C. B. & Long, J. P. Effects of halothane on gastrointestinal motility. Anaesthesiology, 1961, 22, 363-366.

1006. Martin, A., Lathlere, S., Sauvaget, M., Thillier, J. L. & Laget, Y. Enuresis. Electro-gastro-enterographic study. 1st results. Ann. Med. Psychol., Paris, 1973, 2, 279-293.

The E.G.G., thanks to the nocturnal recording of intestinal reflexes, has made it possible to determine the moment of urination under completely physiological conditions. The use of the E.G.G. has confirmed the role of affective factors and of the CNS in the control of the function of the viscera, in particular the bladder, and, knowing the significance of this symptom, to understand the effect of different suggested treatments. The E.G.G. should permit the avoidance of penile examination. Additional work is under way to confirm the observations related in this first study.

1007. Martin, A., Moline, J. & Thouvenot, J. Electrosplanchnographic study of antropyloric evacuation in animals and man. J. Physiol., (Paris), 1969, 1, 151.

The identical results are found in the rat, under analagous experimental conditions. It is thus possible to find the same periodic variations in rhythms of 4 to 5 per minute in connection with the antropyloric activity verified radiologically.

1008. Martin, A. & Thillier, J. L. Electrogastroenterography (E.G.E.G.). Presse Med., 1971, 79, 1235-1237.

Electrogastro-enterography consists of recording from the limbs. This technic is derived from electrosplanchnography. Recording at the level of the abdomen gave inconstant results depending on the position of the electrodes, hence the idea of an electrical axis and the possibility of distant, leads. The advantage of this new method is its great simplicity, its absolute harmlessness and all the possible medical and surgical application such as: aid in diagnosis, postoperative supervision, control of the action of drugs or the effects of conditioned reflexes in psychosomatic medicine.

1009. Martin, A. & Thillier, J. L. Spatiotemporal organization and regulation of digestive motor activities in man: Electro-gastro-enterography (E.G.E.G.). Method of physiological analysis. J. Physiol., Paris, 1971, 63, 253 A.

The article reviews a method of registering the

activity of the stomach small intestine and the colon from the surface of the abdomen, and compares the spatio-temporal organization of digestive, motoric activity in children, adults, and the aged.

1010. Martin, A., Thillier, J. L. & Moline, J. Effects of section of the vagus nerve on gastric motoricity in man. Electrogastroenterographic study. J. Physiol., Paris, 1972, 65, 145A.

The effects of sectioning the vagus nerve are thought to be more of a secretory than a motoric nature, such that 'gastro-plegy' is no more than a consequence of a struggle against functional stenosis of the pylorus.

1011. Martin, A., Thillier, J. L., Moline, J. & Murat, J. Physiological and physiopathological study of digestive motility in humans after vagotomy: Electro-gastro-enterographic control. C. R. Soc. Biol., Paris, 1971, 165, 1967-1970.

Electrogastroenterography demonstrated the efficacy of subphrenic vagotomy in the conscious subject and its effects. It is possible in this way to study the reciprocal influence of the pneumogastric nerve and other autonomic innervations, particularly on the pylorus and stomach. The small intestine does not appear to be affected by sectioning of the pneumogastric and reacts perfectly to distension. The colic reactions are attenuated as a whole but the vesicocolic reflex is maintained intact.

1012. Martin, A., Thouvenot, J. & Touron, P. Periodic changes in abdominal cutaneous potentials in relation to digestive activity. C. R. Soc. Biol., Paris, 1967, 161, 2595-2600.

The recording of cutaneous electrical activity allows for a gathering of global and complex information about the state of the digestive viscera.

1013. Martin, C. L. & Rogers, F. T. Hunger pain. Am. J. Roentgenol. Radium Ther. Nucl. Med., 1927, 17, 222-227.

1014. Martin, J. & Schnedorf, J. G. The absence of changes in gastric activity and of gastrointestinal ulceration following hypothalamic lesions in the monkey and cat. Am. J. Physiol., 1938, 122, 81-85.

1015. Martin, P. D. & Scobie, B. A. Metoclopramide and gastric emptying. N. Z. Med. J., 1968, 67, 494-495.

1016. Martin, W. G. & Morton, H. S. Clinical studies with the electrogastrograph. A.M.A. Arch. Surg., 1952, 65, 382-396.

1017. Martinson, J. The effect of graded stimulation of efferent vagal nerve fibres on gastric motility. Acta Physiol. Scand., 1964, 62, 256-262.

1018. Martinson, J. Studies on the efferent vagal control of the stomach. Acta Physiol. Scand., 1965, 65, Suppl. 255, 1-24.

1019. Martinson, J. The effect of graded vagal stimulation on gastric motility, secretion and blood flow in the cat. Acta Physiol. Scand., 1965, 65, 300-309.

1020. Martinson, J. Vagal relaxation of the stomach. Experimental reinvestigation of the concept of the transmission mechanism. Acta Physiol. Scand., 1965, 64, 453-462.

1021. Martinson, J. & Murin, A. Studies on vagal excitation and inhibition of gastric motility. Acta Physiol. Scand., 1960, 50, Suppl. 175, 103-104.

1022. Martinson, J. & Murin, A. Excitatory and inhibitory effects of vagus stimulation on gastric mobility in the cat. Acta Physiol. Scand., 1963, 57, 309-316.

1023. Matsumoto, H. Periodic potential fluctuations associated with the movement of the fare stomach of goats. Nat. Inst. Anim. Health Quart., 1964, 4, 181-191.

1024. Matsumoto, T., Hayes, M. F., Jr., DeLaurentis, D. & Miyata, M. Evaluation of acupuncture in management of gastrointestinal atony following vagotomy. Surg. Forum, 1972, 23, 401-402.

1025. Maurer, H. J. & Randau, W. Stomach motility and peristalsis following thoractic vago-sympathicosplanchnicotomy. Brus. Beitr. Klin. Chir., 1971, 218, 687-697.

1026. Mautner, H. & Yetwin, I. J. Effects of pentamthylentertrazol (Metrazol) on stomach. J. Am. Pharm. Assoc., 1943, 32, 17-20.

1027. May, W. P. On the movements and innervation of the stomach. Br. Med. J., 1902, 2, 779-781.

1028. Maycock, B. F. & Jones, P. W. A long term remote intragastric pH and motility monitoring system. Biomed. Sci. Instrum., 1969, 5, 127-133.

1029. Mayer, J. The hypothalamic control of gastric hunger contractions as a component of the mechanism of regulation of food intake. Am. J. Clin. Nutr., 1960, 8, 547-561.

1030. Mayer, J. & Sudsaneth, S. Mechanism of hypothalamic control of gastric contractions in the rat. Am. J. Physiol., 1959, 197, 269-280.

1031. McSwiney, B. A. & Wadge, W. J. Effects of variations in intensity and frequency on the contractions of the stomach obtained by stimulation of the vagus nerve. J. Physiol., 1928, 65, 350-356.

1032. Meek, W. J. & Herrin, R. C. The effect of vagotomy on gastric emptying time. Am. J. Physiol., 1934, 109, 221-231.

1033. Meeroff, J. C. The effect of metoclopramide on human gastric emptying and secretion. Acta Gastroenyterol. Latinoam, 1974, 6, 55-61.

1034. Meeroff, J. C., Go, V. L. & Phillips, S. F. The gastric emptying of liquids, measured quantitatively by the arrival of a nonabsorbable indicator in the duodenum. Rev. Esp. Enferm. Apar. Dig., 1973, 41, 131-148.

1035. Meeroff, J. C., Go, V. L. & Phillips, S. F. Gastric emptying of liquids in man. Quantification by duodenal recover marker. Mayo Clin. Proc., 1973, 48, 728-732.

1036. Meeroff, J. C., Go, V. L. & Phillips, S. F. Control of gastric emptying by osmolality of duodenal contents in man. Gastroenterology, 1975, 68, 1144-1151.

1037. Mehnert, H. & Forster, H. Study on the evacuatory mechanism of the stomach after oral administration of different sugars in humans and rats. Diabetologia, 1968, 4, 26-33.

1038. Meigal, Y. A. & Mospanov, L. S. Roentgenometric studies of the motor function of the stomach. Ter. Arkh., 1966, 38, 23-25.

1039. Melchionda, N. & Benfenati, F. Observations on the mathematical elaboration of gastrointestinal manometric recordings. II. Methodological assessment of the quantitative analysis. G. Clin. Med., 1968, 49, 630-645.

1040. Mellinkoff, S. Digestive system. Ann. Rev. Physiol., 1957, 19, 175-204.

1041. Mello, J. B. de., Moreira, A. & Garrido, A., Jr. Vagotomies: Analysis of their position in the treatment of gastroduodenal peptic ulcer. Rev. Hosp. Clin. Fac. Med. Sao Paulo, 1971, 26, 75-82.

1042. Meltzer, S. J. & Auer, J. Vagus reflexes upon esophagus and cardia. Br. Med. J., 1906, 2, 1806-1807.

1043. Mendelson, J. A. & Curtis, G. M. Relationship of gastric motility to abdominal pain in hypothyroidism; preliminary report. Gastroenterology, 1953, 25, 248-272.

1044. Mengrey, R. Gastric inhibitory substance. In C. M. Thompson (Ed.), The Stomach. New York: Grune & Stratton, 1967.

1045. Menguy, R. Role of biliary and pancreatic secretions in the inhibition of gastric motility by fat in the intestine. Am. J. Dig. Dis., 1960, 5, 792-800.

1046. Menguy, R. Motor function of the alimentary tract. Ann. Rev. Physiol., 1964, 26, 227-248.

1047. Meschan, I. & Quigley, J. P. Spontaneous motility of pyloric sphincter and adjacent regions of gut in unanesthetized dog. Am. J. Physiol., 1938, 121, 350-357.

1048. Mescheriakov, F. A. The part of the spinal cord in the regulation of the motor activity of the gastrointestinal tract. Sechenov Physiol. J. USSR, 1959, 11, 83-88.

1049. Mettler, F. G., Spindler, J., Mettler, C. C. & Combs, J. D. Disturbances in gastrointestinal function after localized ablations of cerebral cortex. Arch. Surg., 1936, 32, 618-623.

1050. Meyer, J. Contributions to the physiology of the stomach. XLIX. Hunger and appetite in pulmonary tubercuosis. Arch. Int. Med., 1918, 22, 759-765.

1051. Meyer, J., Sorter, H. & Nicheles, H. Observations on anorexia. Gastroenterology, 1945, 5, 283-289.

1052. Meyer-Burgdorff, H. & Schmidt, W. Der Operierte Magen. Leipzig: Georg Thieme, 1930.

1053. van Miert, A. S. & De la Parra, D. A. Inhibition of gastric emptying by endotoxin (bacterial lipopolysaccharide) in conscious rats and modification of this response by drugs affecting the autonomic nervous system. Arch. Int. Pharmacodyn. Ther., 1970, 184, 27-33.

1054. van Miert, A. S. & van Duin, C. T. The effects of bacterial pyrogens and leucocytic pyrogen upon gastric motility and heart rate frequency in conscious goats. Zentralbl. Veteringermed., 1974, [A], 21, 692-702.

1055. Milenov, K. Electrical and motor activities of the stomach after partial resections of the small and big curvature. Jzv. Mst. Fiziol., (Sofiya), 1968, 11, 79.

1056. Milenov, K. On the rhythm of the electrical and motor activities in intact stomachs and after transverse resections. Izv. Inst. Fiziol., (Sofiya), 1968, 11, 63.

1057. Miolan, J. P. & Roman, C. Modification de l'electromyogramme gastrique du chien par stimulation des nerfs extrinseques. J. Physiol. (Paris), 1971, 63, 561-576.

1058. Miolan, J. P. & Roman, C. Unit discharge of efferent vagal nerve fibers during receptive relaxation of the dog stomach. J. Physiol., Paris, 1974, 68, 692-704.

1059. Mishin, L. N. Teknika elektrogastrografii. (Technique of electrogastrography.) Elektronika v Meditsin, Moscow, 1960.

1060. Misiewicz, J. J. Measurement of intraluminal pressures: Radiotelemetry, design of manometric studies, and computer analysis of records. Am. J. Dig. Dis., 1968, 13, 389-396.

1061. Misiewicz, J. J. The application of motility studies to gastroenterological problems. Postgrad. Med. J., 1968, 44, 749-754.

1062. Misiewicz, J. J., Holdstock, D. J. & Waller, S. L. Motor responses of the human alimentary tract to near-maximal infusions of pentagastrin. Gut, 1967, 8, 463-469.

1063. Misiewicz, J. J., Waller, S. L. & Eisner, M. Motor responses of human gastrointestinal tract to 5-hydroxytryptamine in vivo and in vitro. Gut, 1966, 7, 208-216.

1064. Misiewicz, J. J., Waller, S. L. & Holdstock, D. J. Gastrointestinal motility and gastric secretion during intravenous infusions of gastrin II. Gut, 1969, 10, 723-729.

1065. Misumi, A., Sakamoto, S. & Miyamoto, Y. Electrical potentials of the smooth muscles of the digestive tract-- methods of determination and changes of the results under various conditions. Jap. J. Smooth Muscle Res., Sept., 1974, 10, 233-234.

1066. Mitchell, G. A. G. A microscopic study of the nerve supply of the stomach. J. Anat., London, 1940, 75, 50-63.

1067. Mitchell, G. A. G. Nerve supply of the gastrointestinal tract. Ciba. Clin. Sym., 1961, 13, 39-65.

1068. Miwa, T., Tani, R., Suzuki, S., Fukutomi, H. & Sakita, T. Gastric juice secretion, gastric emptying and its stimulation. Clin. Endocrinol., Tokyo, 1975, 23, 155-159.

1069. Miyajima, Y., Tsugu, Y. & Tatsu, R. Action potentials of the stomach during vagotomy. Jap. J. Smooth Muscle Res., Sept., 1974, 10, 232.

1070. Miyasato, F. Gastroduodenal motility as revealed by the electrogram of smooth muscle. Jiekei. Med. J., 1962, 9, 221-224.

1071. Mkrtichian, S. M. Effect of electro-sleep on the motility of the gastrointestinal tract. Bh. Eksp. Clin. Med., 1975, 15, 84-88.

1072. Moberg, S. Gastric emptying and duodenal digestion before and after partial gastrectomy and selective proximal vagotomy. Scand. J. Gastroenterol., 1974, 24, 3-39.

1073. Moberg, S. & Carlberger, G. Gastric emptying in healthy subjects and patients with various malabsorptive states. Scand. J. Gastroenterol., 1974, 9, 17-21.

1074. Moberg, S. & Carlberger, G. The effect on gastric emptying of test meals with various fat and osmolar concentrations. Scand. J. Gastroenterol., 1974, 9, 29-32.

1075. Moberg, S., Carlberger, G. & Barany, F. Digestion and absorption in the duodenum in relation to gastric emptying in a patient with duodenal ulcer; methodological and physiological considerations. Scand. J. Gastroenterol., 1974, 9, 9-15.

1076. Moberg, S., Carlberger, G. & Barany, F. Digestion and absorption in the duodenum in relation to gastric emptying in patients with peptic ulcer before and after partial gastrectomy or selective proximal vagotomy. Scand. J. Gastroenterol., 1974, 9, 23-28.

1077. Monges, H. Study concerning intra-esophageal and intragastric pressures in normal subjects. Gastroenterologia, (Basel), 1960, 93, 46-55.

1078. Monges, H. & Salducci, J. A method of recording the gastric electrical activity in man. Am. J. Dig. Dis., 1970, 15, 271-276.

1079. Monges, H. & Salducci, J. Electrical activity of the gastric antrum in normal human subjects. Am. J. Dig. Dis., 1971, 16, 623-627.

1080. Monges, H. & Salducci, J. Gastric electromyography. Arch. Fr. Mal. App. Dig., 1973, 62, 419-432.

1081. Monges, H., Salducci, J. & Muller, G. Electromyographic study of the gastric antrum in normal men (preliminary note). Mars. Med., 1968, 105, 861-865.

1082. Monges, H., Salducci, J. & Roman, C. Electromyographic study of normal human gastrointestinal motility. Arch. Fr. Mal. App. Dig., 1969, 58, 517-530.

1083. Morat, J. P. Sur l'innervation motrice de l'estomac. Lyon med., 1882, 40, 289-296; 335-341.

1084. Morat, J. P. Sur quelques particulartes de l'innervation motice de l'estomac et de l'intestin. Arch. de physiol. norm. et path., Paris, 1893, 5, 142-153.

1085. Morgan, E. H. Studies of intraluminal pressure and pH at the gastroesophageal junction. Quart. Bull. Northwestern Univ. Med. School, 1962, 36, 258-267.

1086. Morgan, R. G. H. & Simmons, W. J. The relative effect of diversion of bile to the ileum and to the urinary bladder on fat absorption and on gastrointestinal motility in the rat. Quart. J. Exp. Physiol., 1962, 47, 352-359.

1087. Morishima, K. & Fujitani, J. Zur Untersuchungsmethode der spontanen Bewegung des Froschmagens. Arch. f. exper. Path., 1908, Suppl. B, 407-414.

1088. Morlock, C. G., Hightower, N. C., Code, C. F. & Craig, W. M. Effect of thoracolumbar sympathectomy and splanchnicectomy on antral gastric motility in man. Gastroenterology, 1950, 16, 117-121.

1089. Morris, B. & Morris, R. The effects of cortisone acetate on stomach evacuation and the absorption of 125I-labelled globulins in young rats. J. Physiol., London, 1974, 240, 79-89.

1090. Morrison, A. R. & Habel, R. E. A quantitative study of the distribution of vagal nerve endings in the myenteric plexus of the ruminant stomach. J. Comp. Neurol., 1964, 122, 297-309.

1091. Mortensson, W. Emptying of the stomach after Billroth II resection. Radiologe, 1969, 9, 205-208.

1092. Morton, H. S. The potentialities of the electrogastrograph. Ann. Roy Coll. Surg., England, 1954, 15, 351-373.

1093. Morton, H. S. & Martin, W. S. The electrogastrograph. Some clinical applications. Rev. Gastroenterol., 1958, 20, 37.

1094. Mosin, V. I. & Ter-Akopova, I. G. The functional state of the parathyroid glands and the effect of parathormone on the secretory and motor function of the stomach in peptic ulcer. Ter. Arkh., 1971, 43, 24-26.

1095. Mulinos, M. G. Gastro-intestinal motor response to vagus stimulation after nicotine. Am. J. Anat., 1927, 25, 49-53.

1096. Mulinos, M. G. Gastric hunger mechanism; influence of experimental alterations in blood sugar concentration of gastric hunger contractions. Am. J. Physiol., 1933, 104, 371-378.

1097. Murat, J., Bertrand, J., Crassas, Y., Martin, A. & Thillier, J. L. Early postoperative control of vagotomy by electrogastroenterography. Chirurgie, 1973, 99, 701-710.

1098. Murin, A. Gastric motility after vagotomy in dogs. Acta Chir. Scand., 1956, 112, 98-106.

1099. Murin, A. Influence of the vagal innervation on gastric motor responses to adrenaline and noradrenaline. Acta Physiol. Scand., 1956, 39, 195-202.

1100. Murin, A. Gastric motor responses to adrenaline and noradrenaline after treatment with a parasympathioclytic agent. Acta Physiol. Scand., 1957, 39, 188-194.

1101. Murin, A. Influence of the vagal innervation on gastric motor response to adrenaline and noradrenaline. Acta Physiol. Scand., 1957, 39, 195-202.

1102. Murin, A. The effect of vagotomy on gastric motor responses to drugs in dogs. Acta Physiol. Scand., 1957, 38, 398-414.

1103. Murlaga, S. E. Disorders of the evacuatory function of the resected stomach--A result of noncompensated gastroplegia and acute dilatation of the gastric stump. Vestn. Khir., 1970, 105, 37-39.

1104. Nachlas, M. M., Younis, M. T., Roda, C. P. & Wityk, J. J. Gastrointestinal motility studies as a guide to postoperative management. Ann. Surg., 1972, 175, 510-522.

1105. Nagai, T. & Prosser, C. L. Electrical parameters of smooth muscle cells. Am. J. Physiol., 1963, 204, 915-924.

1106. Nagai, T. & Prosser, C. L. Patterns of conduction in smooth muscle. Am. J. Physiol., 1963, 204, 910-914.

1107. Nagao, F., Aoki, T., Tanaka, N., Yamaguchi, Y. & Ikeuchi, J. Operative policy in gastroduodenal ulcer: Effects of selective proximal or selective vagotomy on gastric secretion and stomach movement. Jap. J. Surg., 1972, 73, 762-764.

1108. Nagaoka, K. Electromyographic study on the mechanism of delayed gastric emptying after vagotomy in dogs. Tohoku. J. Exp. Med., 1968, 95, 1.

1109. Nagler, R. & Spiro, H. M. Persistent gastroesophageal reflux induced during prolonged gastric intubation. New Engl. J. Med., 1963, 269, 495-500.

1110. Nagorna-Stasiak, B. Effect of neurohormones on motor function of the digestive tract in domestic animals. Acta Physiol. Pol., 1968, 19, 225-236.

1111. Nakayama, S. Movement of the digestive tract. Jap. J. Smooth Muscle Res., 1974, 10, 128-131.

1112. Nakayama, S. & Fukuda, H. Effect of cholecystokinin preparations on the movements of the stomach and small intestine. Jap. J. Physiol., 1966, 16, 185-193.

1113. Nakayama, S. & Mori, T. Effects of distention of the gallbladder and bile ducts on the movement of the stomach and intestine of the dog. Jap. J. Physiol., 1967, 17, 458-465.

1114. Nakayama, S. & Mori, T. Effects of stimulation of the phrenic nerve on the movements of the stomach and small intestine. Jap. J. Physiol., 1968, 18, 373-379.

1115. Nakazato, Y. & Ohga, A. Acute inhibitory effects of reserpine on the sympathetically induced response of dog stomach. Jap. J. Pharmacol., 1972, 22, 167-174.

1116. Nakazato, Y., Saito, K. & Ohga, A. Gastric motor and inhibitor response to stimulation of the sympathetic nerve in the dog. Jap. J. Pharmacol., 1970, 20, 131-141.

1117. Nasset, E. S. Digestive system. Ann. Rev. Physiol., 1951, 13, 115-132.

1118. Necheles, H., Levitsky, P., Kohn, R. & Maskin, M. The reaction of the dogs stomach to acetlycholine. Am. J. Physiol., 1936, 116, 112-113.

1119. Necheles, H., Levitsky, P., Kohn, R., Maskin, M. & Frank, R. The vasomotor effect of acetlycholine on the stomach of the dog. Am. J. Physiol., 1936, 116, 330-336.

1120. Needham, D. M. & Shoenberg, C. F. Proteins of the contractile mechanism of mammalian smooth muscle and their possible location in the cell. Proc. Royal Soc. Med., London, 1964, 148, 517-524.

1121. Nelsen, T. S. Use of phaselock techniques for retrieval of the electrogastrogram from cutaneous and swallowed electrodes. Digest of the Seventh International Conference on Medical, Electr. and Biol. Eng., Stockholm, 1967, 23.

Reliable recording of the electrical signal from the stomach of man is made difficult by the low strength of the signal and its contamination by "noise" from skeletal muscle, hair, and gut. Animal experiments have shown the gastric electrical signal is an indicator of certain physiologic states of the stomach and points to its potential usefulness as a specific diagnostic tool in certain human disorders (e.g. emptying disorders).

1122. Nelsen, T. S. A theory of integrated gastrointestinal motor activity based on the chain oscillator model. Am. J. Dig. Dis., 1971, 16, 543-547.

1123. Nelsen, T. S., Bunker, C. E. & Johnson, L. Alterations in muscular and electrical activity of the stomach following vagotomy. Arch. Surg., 1966, 92, 379.

1124. Nelsen, T. S., Eigenbrodt, E. H. & Keoshian, L. A. Motor responses of the canine stomach to insulin and feeding. Arch. Surg., 1966, 92, 379-385.

1125. Nelsen, T. S., Eigenbrodt, E. H., Keoshian, L. A., Bunker, C. E. & Johnson, L. Alterations in muscular and electrical activity of the stomach following vagotomy. Arch. Surg., 1967, 94, 821.

1126. Nelsen, T. S. & Kohatsu, S. Clinical electrogastrography and its relationship to gastric surgery. Am. J. Surg., 1968, 116, 215-222.

Frequently cutaneous abdominal electrodes were used in parallel with the implanted electrodes; this permitted comparison of traces with subsequent improvement of the electronic technics used for retrieval of the gastric electrical signal from the cutaneous recordings.

1127. Nelsen, T. S. & Kohatsu, S. The stomach as a pump. Rendic. R. Gastroenterol., 1971, 3, 65-71.

1128. Nelsen, T. S. & Wallace, J. M. An instrumentation system for on-line processing and statistical analysis of electroenterographic and electrogastrographic data. In: Intern. Conf. Med. Electronics and Biol. Eng. 6th Tokyo, 1965. Jap. Soc. Med. Electr. Biol. Eng., 1965, 369-370.

1129. Nelson, R. S. Movements of the gastric antrum. Gastroenterology, 1956, 30, 669-675.

1130. Nguyen-man-Bihn. The effect of mediators on the motor activity of the multichamber stomach. Nerv. Sist., 1969, 10, 51-56.

1131. Nicolaidis, S. Early systemic responses to orogastric stimulation in the regulation of food and water balance functional and electrophysiological data. Ann. N.Y. Acad. Sci., 1969, 2, 1176-1200.

1132. Nikolov, N. A. & Deleva, Zh. I. Effect of clucagon on the motor and electrical activity of the dog small intestine and stomach. Fiziol. Zh. S.S.S.R., 1975, 61, 774-777.

1133. Nilsson, F. Effect of truncal vagotomy and pyloroplasty on gastric evacuation and small bowel propulsion in the rat. Acta Chir. Scand., 1973, 139, 731-734.

1134. Nilsson, F. & Jung, B. Gastric evacuation and small bowel propulsion after laparotomy. A study with a double isotope technique in rat. Acta Chir. Scand., 1973, 139, 724-730.

1135. Nilsson, F., Jung, B. & Lundquist, H. Gastric evacuation and small bowel propulsion measured with a double isotope technique. A methodological study in the rat. Acta Chir. Scand., 1973, 139, 716-723.

1136. Nimeh, W. Emptying of normal stomach in altitudes. Rev. Gastroenterol., 1947, 15, 763-775.

1137. Nimmo, W. S., Wilson, J. & Prescott, L. F. Narcotic analgesics and delayed gastric emptying during labour. Lancet, 1975, 1, 890-893.

1138. Nimmo, W. S., Wilson, J. & Prescott, L. F. Proceedings: Narcotic analgesics and delayed gastric emptying during labour. Anaesthesiology, 1975, 30, 119.

1139. Nishii, H. & Kuwashima, T. Selective proximal vagotomy and gastric motility. Jap. J. Smooth Muscle Res., Sept., 1974, 10, 159-162.

1140. Nolf, P. Influence des nerfs sympathiques sur la motricite de l'estomac musculaire de l'oiseau. Compt. Rend. Soc. Biol., 1925, 93, 839.

1141. Nolf, P. Du role des nerfs vagues et sympathiques dans l'innervation motrice de l'estomac de l'oiseau. Arch. Int. Physiol., 1927, 28, 309-428.

1142. Northrup, D. W. & Van Liere, E. J. The influence of glucose on gastric emptying. Fed. Proc., 1945, 4, 54.

1143. Northrup, D. W. & Van Liere, E. J. The effect of chloral hydrate on the gastric emptying time in man. Fed. Proc., 1946, 5, 77.

1144. Northrup, D. W. & Van Liere, E. J. Effect of central nervous system depressants on emptying time of human stomach. J. Pharmacol. Exp. Therap., 1948, 93, 208-209.

1145. Northrup, D. W. & Van Liere, E. J. Effect of isomers of amphetamine and desoxyephedrine on gastric emptying in man. J. Pharmacol. Exp. Therap., 1953, 109, 358-360.

1146. Notelovitz, M. Letter: Protein calorie malnutrition, gastric motility and diabetes. S. Afr. Med. J., 1974, 48, 940.

1147. Nozaki, S., Suzuki, K., Tsuchida, H., Abo, M. & Sugiyama, Y. Electromyographic and cinefluorographic study on motility of the residual stomach after gastrectomies. Jap. J. Smooth Muscle Res., 1969, 5, 9-18.

1148. Nylander, B. & Andersson, S. Effect of two methylated prostaglandin E2 analogs on gastroduodenal pressure in man. Scand. J. Gastroenterol., 1975, 10, 91-95.

1149. Nylander, B. & Mattsson, O. Effect of 16, 16-dimethyl PG E2 on gastric emptying and intestinal transit of a barium-food test meal in man. Scand. J. Gastroenterol., 1975, 10, 289-292.

1150. Oashi, S., Omuro, S., Sugawara, I., Kuwata, K. & Okamoto, E. Effects of prostaglandin E1 on canine gastrointestinal motility in vivo and human isolated smooth muscle in vitro. Jap. J. Smooth Muscle Res., 1973, 9, 69-77.

1151. Oberhelman, H. A., Rigler, S. P. & Dragstedt, L. R. Significance of innervation in the function of the gastric antrum. Am. J. Physiol., 1957, 190, 391-395.

1152. Oberhelman, H. A., Rigler, S. P. & Dragstedt, L. R. The significance of innervation in the function of the gastric antrum. Surg. Forum, 1957, 7, 353-356.

1153. Oberhelman, H. A., Woodward, E. R., Zubiran, J. M. & Dragstedt, L. R. Physiology of the gastric antrum. Am. J. Physiol., 1952, 169, 738-748.

1154. Obrink, K. J. Digestion. Ann. Rev. Physiol., 1959, 20, 377-404.

1155. Ochia, B. A. Gastric emptying in young pigs. J. Physiol., London, 1973, 233, 467-480.

1156. Ohga, A., Nakazato, Y. & Saito, K. Considerations of the efferent nervous mechanism of the vago-vagal reflex relaxation of the stomach in the dog. Jap. J. Pharmacol., 1970, 20, 116-130.

1157. Ohga, A., Ota, Y. & Nakazato, Y. The movement of the stomach of sheep with special reference to the omasal movement. Jap. J. Vet. Sci., 1965, 27, 151-160.

1158. Ogi, I. P. E., Khoromski, L. N., Kit, O. N. & Kovalchuk, A. A. Preparation of patients for gastric resection depending on the gastric motor and secretory function. Klin. Khir., 1973, 9, 10-13.

1159. Ohkubo, K. Studies on the intrinsic nervous system of the digestive tract. I. The sub-mucuous plexus of guinea-pig. Jap. J. Med. Sci., 1936, 6, 1-20.

1160. Oi, M., Tanaka, Y. & Iwai, Y. Electrogastromyography and its relationship to gastric surgery. Proc. 7th Intern. Conf. Med. Biol. Eng., Stockholm, 1967, 335.

1161. Oigaard, A. The motor-stimulating effect of metoclopramide and pyridostigemine bromide in normal man and laparotomized patients. A combined study of duodenal electric and motor activity. Scand. J. Gastroenterol., 1975, 10, 65-71.

1162. Okruzhko, I. F. Motor-evacuatory function of the resected stomach in patients with peptic ulcer. Khirurgiia, Mosk., 1969, 45, 32-34.

1163. Olbe, L. & Jacobson, B. Intraluminal pressure waves of the stomach in dogs studied by means of radiosondes in stomach with Pavlov pouches. Gastroenterology, 1963, 44, 787-796.

1164. Oleksin, V. I. & Juziv, K. F. Bioelectrical activity of the stomach and small intestine after cholecystectomy. Klin. Khir., 1975, 1, 64-66.

1165. Ono, K. Stomach surgery and gastric motility. Jap. J. Smooth Muscle Res., 1974, 10, 167-170.

1166. Ortmayer, M. Gastric motor activity in patients with peptic ulcer. Arch. Int. Med., 1925, 35, 423-432.

1167. Osa, T. & Kuriyama, H. The membrane properties and decremental conduction of excitation in the fundus of the guinea-pig stomach. Jap. J. Physiol., 1970, 20, 626-639.

1168. Oshika, H. Proceedings. Effect of phenolphthalein on gastrointestinal transit. Jap. J. Pharmacol., 1974, 24, 5.

1169. Ostick, D. G., Green, G., Howe, K., Dymock, I. W. & Cowley, D. J. Proceedings: A simple clinical method for measurement of gastric emptying of solid meals. Br. J. Surg., 1975, 62, 663.

1170. Paintal, A. S. A study of gastric stretch receptors. Their role in the peripheral mechanism of satiation, hunger and thirst. J. Physiol., London, 1954, 126, 255-270.

1171. Paintal, A. S. The response of gastric stretch receptors and certain other abdominal and thoracic vagal receptors to some drugs. J. Physiol., London, 1954, 126, 271-285.

1172. Paintal, A. S. The influence of certain chemical substances in the initiation of sensory discharges in pulmonary and gastric stretch receptors and atrial receptors. J. Physiol., London, 1957, 135, 486-510.

1173. Pajares, G. J. M., Rodriguez, C. J. & Cano Munoz, R. Motility of the operated stomach. Its cine-endoscopic evaluation and its clinical significance. Rev. Esp. Enferm. Apar. Dig., 1974, 43, 487-494.

1174. Palmer, E. D. Clin. Gastroenterol., (2nd Ed.). New York: Hoeber, 1963.

1175. Palmer, W. L. The mechanism of pain in gastric and in duodenal ulcer. III. The role of peristalsis and spasm. Arch. Int. Med., 1927, 39, 109-133.

1176. Palugyay, J. Rontgenologische Beobachtungen uber die Anatomie und Physiologie der Kardia. Arch. f. d. ges. physiol., 1921, 187, 232-242.

1177. Papazova, M. An Electrophysiological Study of the Motor Activity of the Stomach. Sofia: Publishing House of the Bulgarian Academy of Sciences., 1970.

1178. Papazova, M., Atanassova, E. & Boev, K. On the contractile activity of the longitudinal and circular muscular layers of the stomach. Bull. Inst. Physiol. Bulg. Acad. Sci., 1969, 12, 103-113.

1179. Papazova, M., Atanassova, E. & Boev, K. On the contractile activity of the longitudinal circular muscular layers of the stomach. Izv. Inst. Fiziol., (Sofiia), 1969, 12, 103-114.

1180. Papazova, M., Atanassova, E. & Radovea, A. Correlations of the bioelectric and motor activities between the stomach and duodenum of cats. Compt. Rend. Acad. Bulgare Sci., 1967, 20, 975-977.

1181. Papazova, M., Boev, K., Milenov, K. & Atanassova, E. Mechanical and bioelectrical activity of the stomach wall. Izv. Inst. Fiziol., (Sofiia), 1966, 10, 15-24.

1182. Papazova, M. & Milenov, K. Relationship between the muscle potentials from the stomach and peristalsis. Compt. Rend. Acad. Bulgare Sci., 1964, 17, 523-526.

1183. Papazova, M. & Milenov, K. A method for recording biopotentials from the muscles of the stomach in cats with chronic implanted electrodes. Izv. Inst. Fiziol., (Sofiia), 1965, 9, 187-191.

1184. Papazova, M. & Milenov, K. Relation between spike potentials of electrogastromyogram and stomach peristalsis. Dokl. Bolg. Akad. Nauk., 1965, 18, 585-588.

1185. Papazova, M., Nagai, T. & Prosser, C. L. Two-component slow waves in smooth muscle of cat stomach. Am. J. Physiol., 1968, 214, 695-702.

1186. Pathak, J. D. Some effects of introducing NCl into human stomach. Indian J. Med. Res., 1959, 47, 325-328.

1187. Paton, W. D. M. & Vane, J. R. An analysis of the response of the isolated stomach to electrical stimulation and to drugs. J. Physiol., London, 1963, 165, 10-46.

1188. Patterson, T. L. Vagus and splanchnic influence on the gastric hunger movements of the frog. Am. J. Physiol., 1920, 53, 293-306.

1189. Patterson, T. L. Contributions to the physiology of the stomach. XIII. The variations in the hunger contractions of the empty stomach with age. Am. J. Physiol., 1914, 33, 423-429.

1190. Patterson, T. L. Contributions to the physiology of the stomach. XXIII. The cause of the variations in the gastric hunger contractions with age. Am. J. Physiol., 1915, 37, 316-329.

1191. Patterson, T. L. Contributions to the physiology of the stomach, XXXVI. The physiology of the gastric hunger contractions in the amphibia and the reptilia. Comparative studies I. Am. J. Physiol., 1916, 42, 56-88.

1192. Patterson, T. L. A satisfactory method of demonstrating gastric movements on a small animal. Comparative studies II. J. Lab. Clin. Med., 1920, 5, 674-678.

1193. Patterson, T. L. Gastric tonus of the empty stomach of the frog. Am. J. Physiol., 1920, 54, 153-165.

1194. Patterson, T. L. Gastric tonus of the empty stomach of the frog. Comparative studies IV. Am. J. Physiol., 1920, 54, 153-165.

1195. Patterson, T. L. Vagus and splanchnic influence on the gastric hunger movements of the frog. Comparative studies III. Am. J. Physiol., 1920, 53, 293-306.

1196. Patterson, T. L. The role of the vagi on gastric motility in Necturus maculatus. Proc., Iowa Acad. Sci., 1923, 30, 185-187.

1197. Patterson, T. L. Gastric movements in the pigeon with economy of animal material. Comparative studies V. J. Lab. Clin. Med., 1927, 12, 1003-1008.

1198. Patterson, T. L. The influence of the vagi on the motility of the empty stomach in Necturus. Comparative studies VI. Am. J. Physiol., 1928, 84, 631-640.

1199. Patterson, T. L. Tonic automatism in the stomach of the monkey as the determining factor in the type of its muscular response. Ann. N.Y. Acad. Sci., 1930, 32, 53-86.

1200. Patterson, T. L. Comparative physiology of the gastric hunger mechanism. Ann. N.Y. Acad. Sci., 1933, 34, 55-272.

1201. Patterson, T. L., Illenden, J., Rubright, L. W. & Scott, R. J. The motile state of the stomach of the monkey in hunger. Am. J. Physiol., 1923, 58, 421.

1202. Patterson, T. L. & Rubright, L. W. The influence of tonal conditions on the muscular response of the monkey's stomach. Quart. J. Exp. Physiol., 1934, 24, 3-121.

1203. Patterson, T. L. & Sandweiss, D. J. Relationship between gastroduodenal motility phases and symptoms associated with duodenal ulcer in the human. Am. J. Dig. Dis., 1942, 9, 375-381.

1204. Patyra, S., Koper, B., Kostyra, J. & Koper, S. The influence of vagotomy on the motor activity of rumen and stomach in sheep. Ann. Univ. Mariae Curie Sklodowska., 1972, 27, 103-113.

1205. Paul, F. Comparative studies on pancreas glucagon and secretion effect on motility of proximal and distal human gastrointestinal tract. Acta Microbiol. Pol., 1974, [A], 6, 841-843.

1206. Paul, F. Quantitative studies of the effect of pancreatic glucagon + and secretin ++ on gastrointestinal motility in man recorded by simultaneous electro-manometric registration. Klin. Wochenschr., 1974, 52, 983-989.

1207. Paul, F., Neumann, F. & Schmidt, W. Effect of synthetic salmon calcitonin (SMC) on acid secretion and motility of the human stomach. Verh. Dtsch. Ges. Inn. Med., 1974, 80, 526-528.

1208. Paul, H. E. & Paul, M. F. The effect of mineral oil upon fecal consistency, fecal balls and velocity of gastrointestinal transport. Am. J. Dig. Dis., 1950, 17, 125-129.

1209. Pearcy, J. F. & Van Liere, E. J. Studies on the visceral nervous system. XVII. Reflexes from the colon. I. Reflexes to the stomach. Am. J. Physiol., 1926, 78, 73-74.

1210. Penick, S. B., Smith, G. P. & Hinkle, L. E. Gastric motility in hungry men recorded by a transensor capsule. Clin. Res., 1962, 10, 192.

1211. Penick, S. B., Smith, G. P., Wieneke, K., Jr. & Hinkle, L. E. An experimental evaluation of the relationship between hunger and gastric motility. Am. J. Physiol., 1963, 205, 421-426.

1212. Pera, M. C. & Pera, J. C. Historical data of gastric physiology. Its current status and physiological aspects of the operated stomach. Rev. Esp. Enferm. Apar. Dig., 1973, 41, 331-350.

1213. Perez-Avila, C., Howlett, P. J., Barber, D. C., Sheiner, H. J. & Duthie, H. L. Proceedings: Postoperative symptoms and gastric emptying of solid foods. Gut, 1975, 16, 394-395.

1214. Perren, S. M., Schultheiss, H. R., Meine, J. & Allgower, M. Electrical stimulation types in vagus-nerve-induced stomach contractions. Helv. Chir. Acta., 1970, 37, 250-258.

1215. Perret, G. E. & Hesser, F. H. Studies on gastric motility in the cat. I. Effect of superdiaphragmatic vagotomy, cervical cord transection, sympatho-splanchnicectom and posterior root ganglionectomy. _Gastroenterology_, 1960, _38_, 219-230.

1216. Persson, J. E., Christoffersson, E., Wallenstein, S., Wickbom, G. & Vikterlof, K. J. Studies of stomach emptying using 51 Cr-labelled food. _Nord. Med._, 1971, _86_, 883-884.

1217. Pert, J. H., Davidson, M., Almy, T. P. & Sleisinger, M. H. Esophageal catheterization studies. I. The mechanism of swallowing in normal subjects with particular reference to the vestibule (esophago-gastric sphincter). _J. Clin. Invest._, 1959, _98_, 337-486.

1218. Peskin, G. W. & Miller, L. D. The role of serotonin as an antralinhibitory substance. _Surg. Forum_, 1962, _13_, 245-247.

1219. Peterson, C. G. & Peterson, D. R. Dibutoline, effect of insulin-induced gastric hyper-motility in human subjects and other observations. _Gastroenterology_, 1945, _5_, 169-174.

1220. Petukhov, I. A., Fomchenko, A. I. & Pitkevich, E. S. Electrogastrophy in children with peritonitis. _Khirurgiia_, Mosk., 1972, _48_, 66-69.

Electrogastrograms were recorded in 31 patients with acute appendicitis in order to assess the motor function of the stomach during peritonitis. In 16 patients peritonitis occurred. Eight patients with uncomplicated hernia served as the control. Experiments on dogs were also performed. The ages of the patients ranged between 4 and 14 years. Before the operation the patients with acute appendicitis showed a decrease in voltage of the gastric waves, but after the operation the waves became higher, especially in patients with peritonits. In experimental peritontis a decrease of the mean voltage of the gastric waves was registered with subsequent normalization after treatment. Accordingly, electrogastrography could be used in surgical clinics as method of control of the motor function of the stomach in patients with peritonitis.

1221. Pirk, F. & Skala, I. Roentgenological follow-up of gastrointestinal motility in the course of digestion. _Rev. Czech. Med._, 1972, _18_, 97-100.

1222. Pisarevskii, A. A., Golland, V. S., Chechulin, I. S., Belousov, A. S. & Sokolov, I. N. Device for induction electrostimulation of the gastro-intestinal anastomosis after stomach resection. Med. Tekh., 1973, 6, 49-52.

1223. Pisarevskii, A. A., Golland, V. S., Chechulin, I. S., Belousov, A. S., Sokolov, I. N., Kuznetsov, V. A., Ivanchenko, Y. B., Ovchinnikov, V. I., Bredikhina, N. A. & Shniger, E. V. Device for induction electrostimulation of a gastrointestinal anastamosis after gastric resection. Biomed. Eng., New York, 1974, 7, 374-377.

1224. Plummer, A., Barrett, W. & Rutledge, R. The action of reserpine on the motility of the digestive tract. Am. J. Dig. Dis., 1955, 22, 337-342.

1225. Poensgen, E. Die motorischen Verrichtungen des menschlichen Magens und ihre Storungen mit Ausschluss der Lehre vom Erbrechen. Strassburg, 1882.

1226. Pomosov, D. B. & Iartsev, S. G. Study of the motor evacuatory function of the gastrointestinal tract after resection of the stomach by repairing a segment of the large intestine. Khirurgiia, Mosk., 1966, 42, 129-131.

1227. Porrua, J. M. L., Castro-Rial, M. & Salamanca, F. E., Jr. Accion del A.C.T.H. sobre el functionalismo gastrico en perros hipofisectomizados. II communicacion. Arch. Med. Exptl., Madrid, 1954, 17, 477-502.

1228. Poser, G. & Mertz, D. P. Diagnosis and therapy of gastric secretion and motility disorders. (At the same time, a contribution to the study of secretory capability of the gastric mucosa during the action of vasoactive substances). Verh. Dtsch. Ges. Inn. Med., 1967, 73, 273-278.

1229. Poskonova, M. A. & Mal'Chikova, L. S. Effect of adenine nucleotides on the electrical and mechanical activity of guinea-pig talnia coli and frog stomach circular muscle. Dok. Akad. Nauk. S.S.S.R., 1970, 190, 1494-1497.

1230. Postlethwait, R. W., Bradshaw, H. H., McRae, J. T., Williams, R. W. & Beaton, W. R., Jr. Gastric secretion and motility after vagotomy in dogs. Gastroenterology, 1950, 15, 320.

1231. Postlethwait, R. W., Hill, H. V., Jr., Chittum, J. R. & Grimson, K. S. Effect of vagotomy and drugs on gastric motility. Ann. Surg., 1948, 128, 184.

1232. Potter, J. M., Sparrow, M. P. & Simmons, W. J. Increased uptake and efflux of calcium with acetylcholine stimulation in smooth muscle of toad stomach. Aust. J. Exp. Biol. Med. Sci., 1970, 48, 429-443.

1233. Poulsen, J. S. & Jones, B. E. The influence of metabolic alkalosis and other factors on the abomasal emptying rates in goats and cows. Nord. Vet. Med., 1974, 26, 22-30.

1234. Prosser, C. L. Conduction in nonstriated muscles. Physiol. Rev., 1962, 42, 193-212.

1235. Prosser, C. L. Smooth muscle. Ann. Rev. Physiol., 1974, 36, 503-535.

1236. Prosser, C. L., Burnstock, G. & Kahn, J. Conduction in smooth muscle; comparative structural properties. Am. J. Physiol., 1960, 199, 545-552.

1237. Prosser, C. L. & Nagai, T. Patterns of conduction in smooth muscle. Am. J. Physiol., 1963, 204, 910-914.

1238. Puskov, N. I. The electrographic method of examining the motor function of the stomach in patients with thermal burns. Klin. Khir., 1967, 10, 51-54.

1239. Quast, D. C., Beall, A. C., Jr. & De Bakey, M. E. Clinical evaluation of the gastrointestinal pacemaker. Surg. Gynecol. Obstet., 1965, 120, 35-37.

1240. Quigley, J. P. A modern explanation of the gastric emptying mechanism. Am. J. Dig. Dis., 1943, 10, 418-421.

1241. Quigley, J. P. Digestive tract: Intralumen pressures with special reference to gastrointestinal propulsion and gastric evacuation. Med. Physics, 1944, 1, 310-318.

1242. Quigley, J. P. Digestive system. Ann. Rev. Physiol., 1946, 8, 145-168.

1243. Quigley, J. P. The role of the digestive tract in regulating the ingestion of food. Ann. N.Y. Acad. Sci., 1955-1956, 63, 6-14.

1244. Quigley, J. P., Barcroft, J., Adair, G. S. & Goodman, E. N. The difference in potential across gastric membranes and certain factors modifying the potential. Am. J. Physiol., 1937, 119, 763.

1245. Quigley, J. P. & Brody, D. A. The motor activity of the pyloric sphincter studied by the pyloric inductograph. Fed. Proc., 1944, 3, 37-38.

1246. Quigley, J. P. & Brody, D. A. Digestive tract: Intralumen pressures; gastrointestinal propulsion, gastric evacuation, pressure-wall tension relationships. In O. Glasser (Ed.), Medical Physics. Chicago: Year Book, 1950.

1247. Quigley, J. P. & Brody, D. A. Physiologic and clinical consideration of pressures developed in digestive tract. Am. J. Med., 1952, 13, 73-81.

1248. Quigley, J. P. & Hallaran, W. R. The independence of spontaneous gastrointestinal motility and blood sugar levels. Am. J. Physiol., 1932, 100, 102-110.

1249. Quigley, J. P., Johnson, V. & Solomon, E. I. Action of insulin on the motility of the gastro-intestinal tract. I. Action on the stomach of normal fasting man. Am. J. Physiol., 1929, 90, 89-98.

1250. Quigley, J. P. & Louckes, H. The effect of complete vagotomy on the pyloric sphincter and the gastric evacuation mechanism. Gastroenterology, 1951, 19, 533-537.

1251. Quigley, J. P. & Louckes, H. Effect of NAHCO on tone and motility of the pyloric sphincter as recorded by the inductograph. Fed. Proc., 1954, 13, 114.

1252. Quigley, J. P. & Louckes, H. Gastric emptying. Am. J. Dig. Dis., 1962, 7, 672-676.

1253. Quigley, J. P. & Meschan, I. The gastric evacuation of fats with especial reference to the pyloric sphincter activity. Rev. Gastroenterol., 1937, 4, 272-275.

1254. Quigley, J. P. & Meschan, I. The role of the vagus in the regulation of the pyloric sphincter and adjacent portions of the gut, with especial reference to the process of gastric evacuation. Am. J. Physiol., 1938, 123, 166.

1255. Quigley, J. P. & Meschan, I. Inhibition of the pyloric sphincter region by the digestion products of fat. Am. J. Physiol., 1941, 134, 803-807.

1256. Quigley, J. P. & Phelps, K. R. The mechanism of gastric motor inhibition from ingested carbohydrate. Am. J. Physiol., 1934, 109, 133-138.

1257. Quigley, J. P., Read, M. R., Radzow, K. H., Meschan, I. & Werle, J. M. The effect of hydrochloric acid on the pyloric sphincter, the adjacent portions of the digestive tract and on the process of gastric evacuation. Am. J. Physiol., 1942, 137, 155-159.

1258. Quigley, J. P. & Templeton, R. D. Action of insulin on the motility of the gastrointestinal tract. Am. J. Physiol., 1930, 91, 475-495.

1259. Quigley, J. P. & Hallaran, W. R. The independence of spontaneous gastro-intestinal motility and blood sugar levels. Am. J. Physiol., 1932, 100, 102-110.

1260. Quigley, J. P. & Templeton, R. D. Action of insulin on the motility of the gastro-intestinal tract. III. a. Action on the pyloric pouch. b. Action on the stomach following double splanchnicotomy. Am. J. Physiol., 1930, 91, 475-481.

1261. Quigley, J. P. & Templeton, R. D. Action of insulin on the motility of the gastro-intestinal tract. IV. Action on the stomach following double vagotomy. Am. J. Physiol., 1930, 91, 482-487.

1262. Quigley, J. P., Werle, J. M. & Brody, D. A. Correlation of antral and bulbar pressures with fluorscopic observations during gastric evacuation. Am. J. Dig. Dis., 1940, 7, 434-435.

1263. Quigley, J. P., Werle, J. M., Ligon, E. W., Jr., Read, M. R., Radzow, K. H. & Meschan, I. The influence of fats on motor activity of the pyloric sphincter region and on the process of gastric evacuation studied by the balloon-water manometer and by the optical manometer-fluoroscopic technics. Am. J. Physiol., 1941, 134, 1132-140.

1264. Quigley, J. P., Zettelman, H. J. & Ivy, A. C. Analysis of the factors involved in gastric motor inhibition by fats. Am. J. Physiol., 1934, 108, 643-651.

1265. Quist, G., Dormandy, J., Brown, C., Slome, D. & Scott, G. B. The experimental production of gastric ulcers by induced muscle spasm. Br. J. Surg., 1974, 61, 259-263.

1266. Rabkin, I. K. H., Svarasov, V. M. & Arablinskii, V. M. The motor function of the discharge end of the stomach. Ter. Arkh., 1968, 40, 85-88.

1267. Racugno, V. & Sannia, A. Toentgencinematographic study of gastric motility in the trendelenburg position. Nut. Radiol., 1967, 33, 329-336.

1268. Radecki, T., Konturek, S. & Kaulbersz, J. Studies on the influence of fat, glucose, saline and peptone solution on gastric motility in rats. Acta Physiol. Pol., 1970, 21, 499-508.

1269. Ralphs, D. N., Bloom, S. R., Lawson-Smith, C. & Thomson, J. Proceedings: The relationship between gastric emptying rate and plasma enteroglucagon concentration. Gut, 1975, 16, 406.

1270. Ramsbottom, N. & Hunt, J. N. Studies of the effect of metoclopramide and apomorphine on gastric emptying and secretion in man. Gut, 1970, 11, 989-993.

1271. Rang, H. P. The kinetics of action of acetylcholine antagonists in smooth muscle. Proc. Royal Soc. Med., 1966, 164, 488-510.

1272. Ranson, S. W. Afferent paths for visceral reflexes. Physiol. Rev., 1921, 1, 477-522.

1273. Ravdin, I. S., Kneisel, J. J., Jr., Taylor, R., Lemon, H. M., Thompson, E. M. & Smithwick, R. H. Measurement of electropotentials of stomach. Surg. Forum, Amer. Col. of Surg., 1950, 69.

1274. Read, R. C. & Hall, W. H. Objective assessment of gastric function after vagotomy. Curr. Probl. Surg., 1974, 1-63.

1275. Rebrov, V. G. Diagnostic possibilities of electrogastrography from parts of the body distant from the stomach. Klin. Med., 1974, 52, 115-120.

1276. von Redwitz, E. F. Die Physiologie des Magens nach Resektion aus der Kontinuitat. Mitt. a. d. Grenzgeb. d. Med. u. Chir., 1917, 29, 531-620.

1277. Reed, J. D. & Sanders, D. J. Pepsin secretion, gastric motility and mucosal blood flow in the anaesthetized cat. J. Physiol., London, 1971, 216, 159-170.

1278. Regan, J. F. The action of insulin on the motility of the empty stomach. Am. J. Physiol., 1933, 104, 90-94.

1279. Rehm, W. S. Positive injury potentials of stomach. Am. J. Physiol., 1944, 140, 720-725.

1280. Rehm, W. S. Evidence that major portion of gastric potential originates betwten submucosa and mucosa. Am. J. Physiol., 1946, 147, 69-77.

1281. Rehm, W. S. & Enelow, A. J. The effect of thiocyanate on gastric potential and secretion. Am. J. Physiol., 1945, 144, 701-705.

1282. Rehner, M., Winkler, R., Soehendra, N. & Dahm, K. Motility disorders following combined surgery of the stomach. Zentralbl. Chir., 1974, 99, 723-729.Z.

1283. Reid, C. S. W. Eructation and rumen movements in sheep. J. Physiol., 1960, 153, 39-40.

1284. Reifferscheid, M. Postoperative disorders of the stomach-intestine activity as a sign of autonomic reactions. Acta Neuroveg., Wien, 1967, 30, 501-511.

1285. Reinke, D. A., Rosenbaum, A. H. & Bennett, D. R. Patterns of dog gastrointestinal contractile activity monitored in vivo with extraluminal force transducers. Am. J. Dig. Dis., 1967, 12, 113-141.

1286. Reynolds, L. & McClure, C. W. Motor phenomena occurring in normal stomaches, in the presence of peptic ulcers and its pain, as observed fluoroscopically. Arch. Int. Med., 1922, 29, 1-11.

1287. Reznik, S. D. & Pesoki, A. N. The effect of the method of restoration of gastrointestinal continuity on the motor and evacuative activity of the restricted stomach. Vestn. Rentgenol. Radiol., 1969, 44, 93-94.

1288. Rhodes, J. & Goodall, P. Apsimon-mechanics of gastroduodenal emptying. Gut, 1966, 7, 515-520.

1289. Rice, H. V. & Ross, R. J. Factors affecting electrical potential of gastric mucosa. Am. J. Physiol., 1947, 149, 77-94.

1290. Richardson, C. T., Bailey, B. A., Walsh, J. H. & Fordtran, J. S. The effect of an H2-receptor antagonist on food-stimulated acid secretion, serum gastrin, and gastric emptying in patients with duodenal ulcers. Comparison with an anticholinergic drug. J. Clin. Invest., 1975, 55, 536-542.

1291. Richardson, K. C. Electron miscroscopic observations on Auerbach's plexus in the rabbit with special reference to the problem of smooth muscle innervation. Am. J. Anat., 1958, 103, 99-136.

1292. Richter, C. P. Action currents from the stomach. Am. J. Physiol., 1924, 67, 612-633.

1293. Riddle, O. The rate of digestion in cold-blooded vertebrates. The influence of season and temperature. Am. J. Physiol., 1909, 24, 447-458.

1294. Riggs, W., Jr. & Long, L. The value of the plain film roentgenogram in pyloric stenosis. Am. J. Roentgenol. Radium Ther. Nucl. Med., 1971, 112, 77-82.

1295. Riley, J. L. & Cook, H. M. A small, inexpensive radio-telemetering capsule for measuring rumen motility. Cornell Vet., 1974, 64, 225-232.

1296. Roberts, W. M. Effect of oils on gastric secretion and motility. Quart. J. Med., 1931, 24, 133-152.

1297. Robertson, P. A. & Rubin, D. An indirect action of angiotension I on smooth muscle. Nature, 1958, 182, 867-868.

1298. Roche, M. & Santini, R. Comparative study of gastric electromyography. J. Physiol., Paris, 1970, 62, Suppl. 3, 441.

1299. Roche, M. & Santini, R. Electrophysiologic study of the gastric motility. I. Comparative electrogastromyography. C. R. Soc. Biol., Paris, 1970, 164, 1528-1534.

1300. Rogers, F. T. & Bergovitz, Z. A note on the role of the intrinsic plexus in determining the effects on gastric motility of vagus stimulation. Am. J. Physiol., 1921, 56, 257-263.

1301. Rhodes, J. & Goodall, P. Apsimon-mechanics of gastroduodenal emptying. Gut, 1966, 7, 515-520.

1302. Rogatz, J. L. Roentgen-ray studies of stomach function. Am. J. Dis. Child., 1924, 28, 53-68.

1303. Rogers, F. T. Contributions to the physiology of the stomach. XX. The contractions of the rabbit's stomach during hunger. Am. J. Physiol., 1915, 36, 183-190.

1304. Rogers, F. T. Contributions to the physiology of the stomach. XXXIX. The hunger mechanism of the pigeon and its relation to the central nervous system. Am. J. Physiol., 1916, 41, 555-570.

1305. Rogers, F. T. Reflex control of gastric vagus-tonus. Am. J. Physiol., 1917, 42, 605-606.

1306. Rogers, F. T. & Bercovitz, Z. A note on the role of the intrinsic plexuses in determining the effects on gastric motility of vagus stimulation. Am. J. Physiol., 1921, 56, 257-263.

1307. Rogers, F. T. & Hardt, L. L. J. Contributions to the physiology of the stomach. XXVI. The relation between the digestion contractions of the filled, and the hunger contractions of the "empty" stomach. Am. J. Physiol., 1915, 38, 274-284.

1308. Rogers, F. T. & Martin, C. L. X-ray observations of hunger contractions in man. Am. J. Physiol., 1926, 76, 349-353.

1309. Rogers, F. T. & Martin, C. L. A roentgenological study of gastric hunger contractions. Am. J. Physiol., 1927, 82, 113.

1310. Rosenbaum, A. H., Reinke, P. A. & Bennett, D. R. In vivo force frequency and velocity of dog gastrointestinal contractile activity. Am. J. Dig. Dis., 1967, 12, 142-153.

1311. Rosenberg, D., Secaf, F., Carvalhon, D. E. & Aratangy, L. R. Gastrointestinal motility in patients with stomach surgery for gastroduodenal ulcer. Rev. Paul. Med., 1969, 75, 243-250.

1312. Rosenbluth, J. Smooth muscle: An ultrastructural basis for the dynamics of its contraction. Science, 1965, 148, 1337-1339.

1313. Rosenthal, N. & Able, H. A. Blood changes following partial or total gastrectomy. Tr. Am. Gastro-enterol. A., 1933, 215-225.

1314. Rosenthal, S. & Nasset, E. S. Gastric emptying and intestinal absorption of carbohydrate and protein as influenced by the nature of the test meal. J. Nutr., 1958, 66, 91-103.

1315. Rossbach, M. J. Beitrage zur Lehre von den Bewegungen des Magens, Pylorus und Duodenums. Deutsch. Arch. f. klin. Med., 1890, 46, 296-322.

1316. Rothnie, N. G., Kemp, H. R. A. & Catchpole, B. N. Early postoperative gastrointestinal activity. Lancet, 1963, 20, 64.

1317. Roux, J. C. & Balthazard, M. Etude des contractions de l'estomac chez l'homme a l'aide des rayons de Rontgen. Compt. rend. de la Soc. de Biol., Paris, 1897, 49, 785-787.

1318. Roux, J. C. & Balthazard, M. Note sur les fonctions motrices de l'estomac du chien. Compt. rend. de la Soc. de Biol., Paris, 1897, 49, 704-706.

1319. Rowlands, E. N. Gastrointestinal motility in man. In F. Avery Jones (Ed.), Modern Trends in Gastro-enterology. New York: Hoeber, 1952.

1320. Rowlands, E. N., Honour, A. J., Edwards, D. A. W. & Corbett, B. D. Metal capsule optical manometer for measuring pressure in the human gut with an open-ended tube. Clin. Sci., 1953, 12, 299-306.

1321. Roze, C., Chariot, J. & Debray, C. Action of ethyl urethane on the gastric evacuation in the rat. C. R. Soc. Biol., Paris, 1970, 164, 1955-1958.

1322. Ruch, T. C. Motility of the gastrointestinal tract. In T. C. Ruch & D. Patton (Eds.), Textbook of Physiology, 19th Ed., Philadelphia: Saunders Co., 1965.

1323. Ruckenbusch, Y. & Tomov, T. The sequential contractions of the rumen associated with eructation in sheep. J. Physiol., 1973, 235, 447-458.

1324. Ruckenbusch, Y. & Laplace, J. P. Comparative study of rumen-reticulum motility during merycic rejection and vomiting in the sheep. C. R. Soc. Biol., Paris, 1966, 160, 2108-2113.

1325. Rupp, A. & Schultz, F. W. Motility of the empty stomach in normal and in malnourished asthenic children. Am. J. Dis. Child., 1930, 39, 241-256.

1326. Ruppin, H., Domschke, S., Domschke, W., Wunsch, E., Jaeger, E. & Demling, L. Effects of 13-nle-motilin in man--inhibition of gastric evacuation and stimulation of pepsin secretion. Scand. J. Gastroenterol., 1975, 10, 199-202.

1327. Russell, R. W. & Stern, R. M. Gastric motility: The electrogastrogram. In P. H. Venables & I. Martin (Eds.), Manual of Psychophysiological Methods. Amsterdam: North Holland, 1967.

> The present chapter describes such a method, one which was first used in our laboratory by Davis and his co-workers (Davis et al., [1957], Davis et al. [1959]). This method of recording the electrogastrogram (EGG), from electrodes fixed to the skin surface of the abdomen has undergone further development in the process of standardizing the detailed procedure described below. The present chapter starts with a brief history of various methods for recording gastric activity; it then proceeds to discuss the nature of the EGG and the techniques for analyzing the records obtained. It ends with a brief summary of research results which provide information about the source of the potentials and about the validity of the method.

1328. Ryle, J. A. Gastric Function in Health and Disease. London: Oxford University Press, 1926.

1329. Saito, S., Toda, N., Matsumoto, M. & Tobe, T. Roentgenologic studies on diseases of the stomach. 5. Changes in contour and peristalsis. Nippon Acta Radiol., 1971, 30, 1102-1118.

1330. Sakamoto, S. Study of the electrical and mechanical activities in stomach smooth muscle: Effects of splanchnic nerve stimulation, catecholamine, reserpine and phentolamine. J. Kumamoto Med. Soc., 1971, 45, 229-239.

1331. Sakamoto, Y. & Kuriyama, H. The relationship between the electrical and mechanical activity of the guinea-pig stomach. Jap. J. Physiol., 1970, 20, 640-656.

1332. Sakamoto, S. & Wishi, K. Electrical and mechanical activities in the stomach of the dog: Effects of the vagus nerve stimulation, atropine, neostigmine and adrenaline. Kumamoto Med. J., 1967, 20, 25-38.

1333. Salamanca, F. E., Jr. La fisiologia gastrica humana seguh les datos de mi procedimento de exploracion functional. Rev. Clin. Espan., 1943, 8, 94-104.

1334. Salamanca, F. E., Jr. Estudio de fisiologia gastrica en el perro. Trabajos Inst. Nacional Cienc. Med., 1949, 12, 2-64.

1335. Salamanca, F. E., Jr. Las technicas de exploracion funcional gastrica aplicadas al estudio de la fisiologia del perro. Medicina, 1950, 18, 2-29.

1336. Salupere, V. P. Paresis of the stomach in diabetes mellitus. Klin. Med., Mosk., 1971, 49, 155-156.

1337. Sangster, W., Grossman, M. I. & Ivy, A. C. Effect of d-amphetamine on gastric hunger contractions and food intake in dogs. Am. J. Physiol., 1948, 153, 259-263.

1338. Sangster, W., Grossman, M. I. & Ivy, A. C. Relation of rate of gastric evacuation to time of onset of gastric hunger contractions. J. Appl. Physiol., 1949, 1, 637-642.

1339. Saredi, R. J., Paterson, G., Musso, A., Sabagh, R. & Mejia, M. M. Gastrointestinal passage and steatorrhea in severe chornic atropic gastritis. Prensa. Med. Argent., 1967, 54, 2108-2110.

1340. Sarna, S. K. Gastrointestinal electrical activity: Terminology. Gastroenterology, 1975, 68, 1631-1635.

1341. Sarna, S. K., Bowes, K. L. & Daniel, E. E. Post-operative gastric electrical control activity (ECA) in man. In E. E. Daniel et al. (Eds.) Proc. 4th International Symposium on Gastrointestinal Motility. Vancouver: Mitchell Press, 1974.

1342. Sarna, S. K. & Daniel, E. E. Electrical stimulation of gastric electrical control activity. Am. J. Physiol., 1973, 225, 125-131.

1343. Sarna, S. K. & Daniel, E. E. Threshold curves and refractoriness properties of gastric relaxation oscillators. Am. J. Physiol., 1974, 226, 749-755.

1344. Sarna, S. K. & Daniel, E. E. Vagal control of gastric electrical control activity and motility. Gastroenterology, 1975, 68, 301-308.

1345. Sarna, S. K., Daniel, E. E. & Kingma, Y. J. Effects of partial cuts on gastric electrical control activity and its computer model. Am. J. Physiol., 1972, 223, 332-340.

1346. Sarna, S. K., Daniel, E. E. & Kingma, Y. J. Premature control potentials in the dog stomach and in the gastric computer model. Am. J. Physiol., 1972, 222, 1518-1523.

1347. Sarna, S. K., Daniel, E. E. & Kingma, Y. J. Simulation of the electric-control activity of the stomach by an array of relaxation oscillators. Am. J. Dig. Dis., 1972, 17, 299.

1348. Sato, A. Electrophysiological studies on the working mechanisms of muscle walls of the stomach. Jap. J. Physiol., 1960, 10, 357-373.

1349. Sato, A., Sato, Y., Shimada, F. & Torigata, Y. Changes in gastric motility produced by nociceptive stimulation of the skin in rats. Brain Res., 1975, 87, 151-159.

1350. Sawyer, P. N., Rhoads, J. E. & Panzer, R. An evaluation of electrogastrography in the diagnosis of gastric cancer. Surgery, 1949, 26, 479.

1351. Scantlebury, R. E. The effect of psychic phenomena on the movements of the empty stomach in man. Wayne University Graduate Student Monographs in Science, 1940, 1, 32.

1352. Scantlebury, R. E. & Patterson, T. L. The effect of certain psychic phenomena on the various phases of gastric hunger contractions. Proceedings Am. J. Physiol., 1938, 123, 179.

1353. Scantlebury, R. E. & Patterson, T. L. Hunger motility in a hypnotized subject. Quart. J. Exp. Physiol., 1940, 30, 347.

1354. Schalk, A. F. & Amadon, R. S. Gastric motility studies in the goat and the horse. J. Am. Vet. Med. Assoc., 1921, 59, 151-172.

1355. Schalk, A. F. & Amadon, R. S. Physiology of the ruminant stomach (bovine). Study of the dynamic factors. North Dakota Agric. Exp. Sta. Bull., 1928, No. 216, 3-64.

1356. Schapiro, H., Cummins, A. J. & Luckey, L. Dissociation between gastric secretion and motility. Am. J. Dig. Dis., 1965, 10, 751-756.

1357. Schapiro, H. & Woodward, E. R. Inhibition of gastric motility by acid in the duodenum. J. Appl. Physiol., 1955, 8, 121-127.

1358. Schapiro, H. & Woodward, E. R. Pathway of enterogastric reflex. Proc. Soc. Exp. Biol. Med., 1959, 101, 407-409.

1359. Scharli, A. F. & Leditschke, J. F. Gastric motility after pyloromyotomy in infants. A reappraisal of postoperative feeding. Surgery, 1968, 64, 1133-1137.

1360. Schell, N. B., Karelitz, S. & Epstein, B. A. Radiographic study of gastric emptying in premature babies. J. Pediat., 1963, 62, 342-347.

1361. Schenk, E. A. & Frederickson, E. L. Cardiac and coicopharyngeal sphincter thresholds in the cat. Am. J. Physiol., 1959, 197, 743-746.

1362. Schenk, E. A. & Frederickson, E. L. Pharmacologic evidence for a cardiac sphincter mechanism in the cat. Gastroenterology, 1961, 40, 75-80.

1363. Schindler, R. Mechanism and treatment of cardiospasm. Munchen Med. Wochenschr., 1926, 73, 1612.

1364. Schlegel, J. F., Coburn, W. M., Jr. & Code, C. F. Gastric emptying of solid and complaint spheres in dogs. Physiologist, 1966, 9, 283.

1365. Schmid, E. & Hintz, G. Gastric motility in duodenal ulcer. Z. Gastroenterol., 1971, 9, 367-370.

1366. Schmid, E., Wagner, T. & Ritter, U. Inhibition of gastric motility by rectal administration of hyoscine-N-butylbromide. Arzneim. Forsch., 1971, 21, 813-815.

1367. Schmidt, A. Uber die beziehungen des saugreflexes zur magentatigkeit. (Relation of the sucking reflex to activity of the stomach). Ztschr. f. Kinderh., 1927, 45, 20-27.

1368. Schmidt, H. On the effect of respiration on the vegetatively guided movement processes of the stomach and the duodenum. Med. Welt., 1967, 47, 2839-2844.

1369. Schofield, B. & Chir, B. The inhibition of pepsin output in separated gastric pouches in dogs following feeding, and its correlation with motility changes. Gastroenterology, 1959, 37, 169-181.

1370. Schofield, G. C. Anatomy of muscular and neural tissues in the alimentary canal. In C. F. Code (Ed.), Handbook of Physiology. Sect. 6, Alimentary Canal. Vol. IV. Motility. Washington, D.C.: American Physiological Society, 1968.

1371. Schrumpf, E. & Stadaas, J. Effect of gastrid distention on motility and plasma gastrin concentration before and after secretin administration. Scand. J. Gastroenterol., 1974, 9, 119-122.

1372. Schulz, J., Reitzig, P., Koblitz, F., Lisewski, G. & Schulze, E. Elektrogastrographische Untersuchungen bei akuten gastroenterologischen Erkrankungen. Deutsches Gesundheitswesen, 1975, 30, 956-958.

The authors report on the applicability of electrogastroenterography in patients suffering from acute illness of the gastric-intestinal tract. It could have been stated, that repeated registrations will bring forth useful results in respect to the evaluation of the gastric-intestinal motility during the course of the illness. In patients with an acute condition of their illness a retardation of the motility could have been observed, that was extensively approaching to normal values after a fading away of acute clinical symptoms.

1373. Schulz, J., Reitzig, P., Schulze, E., Barsch, J., Etzold, H. & Lisewski, G. Uber die Aussagemoglichkeiten der Elektrogastrographie. Deutsches Gesundheitswesen, 1973, 28, 2485-2488.

The results so far permit the conclusion that registration of action potentials of the gastrointestinal tract is possible by means of electrodes affixed to the skin over the stomach wall. In this connection, more information appears to be obtained by means of local rather than peripheral electrodes. The first observations after injecting metropraniol suggest that the procedure may, for example, be applied for the examination of the efficacy of certain drugs. However, it must be emphasized that electrogastroenterography has not yet fully matured despite its relatively long history and further investigations are necessary in order to be able to assess fully their value. That is especiallyy true for the very ambitious expectations by some authors concerning the possibility of early diagnosis of gastrointestinal disorders.

1374. Schutz, E. Therapy of functional gastroduodenal evacuation disorders with metoclopramide (paspertin). Munchen Med. Wochenschr., 1967, 109, 1227-1231.

1375. Seaman, W. B. Motor dysfunction of the gastrointestinal tract. Am. J. Roentgenol. Radium Ther. Nucl. Med., 1972, 116, 235-244.

1376. Semba, T. Studies on the enterogastric reflexes. Hiroshima J. Med. Sci., 1954, 2, 323-328.

1377. Semba, T. Studies on the gastro-colic reflexes. Hiroshima J. Med. Sci., 1954, 2, 329-333.

1378. Semba, T. Motor responses of the stomach and small intestine to stimulation of the urinary bladder. Jap. J. Physiol., 1956, 6, 294-299.

1379. Semba, T., Fujii, I. & Mizonishi, T. Proceedings: Atropine resistant promotion of gastric motility. J. Physiol. Soc. Jap., 1974, 36, 289.

1380. Semba, T., Fujii, K. & Fujii, Y. The responses of gastric motility and their location by stimulating the thoracic cord of the dog. Hiroshima J. Med. Sci., 1970, 19, 73-85.

1381. Semba, T., Fujii, K. & Fujii, Y. Control of gastric motility by the thoracic cord. Brain Nerve, (Tokyo), 1971, 23, 679-684.

1382. Semba, T., Fujii, K. & Kimura, N. The vagal inhibitory response of the stomach to stimulation of the dog's medulla oblongata. No. To. Shinkei., 1965, 17, 485-489.

1383. Semba, T., Fujii, K. & Kimura, N. The vagal inhibitory responses of the stomach to stimulation of dog's medulla oblongata. Jap. J. Physiol., 1964, 14, 319-327.

1384. Semba, T., Kimura, N. & Fujii, K. Bulbar influence on gastric motility. Jap. J. Physiol., 1969, 19, 521-533.

1385. Semba, T. & Hiroka, T. Motor responses of the stomach and small intestine caused by stimulation of the peripheral end of the splanchnic nerve, thoracic sympathetic trunk and spinal roots. Jap. J. Physiol., 1957, 7, 64-71.

1386. Semba, T., Noda, H. & Fujii, K. On splanchnic motor responses of stomach movements produced by stimulation of the medulla oblongata and spinal cord. Jap. J. Physiol., 1963, 13, 466-478.

1387. Serdyuk, E. N. The influence of thiamine on the motor and interoceptive function of the stomach. Boprosy Pitaniya, 1953, 12, 16-20. (Also in Chem. Abstracts, 1953, 47, 12546).

1388. Setekleiv, J. Innervation and motility of the stomach. Tidsskr. Nor. Laegeforen., 1967, 87, 816-819.

1389. Sethbhakdi, S., Sricharoen, V. & Wilairatana, S. The effect of metoclopramide on gastric emptying. Am. J. Proctol., 1975, 26, 51-56.

1390. Seymour, C. & Weinberg, J. A. Emotions and gastric activity. J. Am. Med. Assoc., 171, 1959, 1193.

1391. Shahidullah, M., Kennedy, T. L. & Parks, T. G. The vagus, the duodenal brake, and gastric emptying. Gut, 1975, 16, 331-336.

1392. Shapiro, H. & Stoner, E. K. Gastric temperature adaptation. Am. J. Gastroenterol., 1968, 49, 391-404.

1393. Shapiro, K. L. & Ivy, A. C. The effect of "carbonation" on the rate of gastric emptying and intestinal absorption of sucrose solution. Fed. Proc., 1948, 7, 112.

1394. Sharma, K. N., Anand, B. K., Cua, S. & Singh, B. Role of stomach in regulation of activities of hypothalamic feedings centers. Am. J. Physiol., 1961, 201, 593-598.

1395. Sharma, M. L. & Grewal, R. S. The inhibition of peristaltic reflex by sympathomimetic drugs. Indian J. Physiol. Pharmacol., 1963, 7, 224-234.

1396. Shay, H. Pathologic physiology of gastric and duodenal ulcer. Bull. N.Y. Acad. Med., 1944, 20, 264-291.

1397. Shay, H. & Gershon-Cohen, R. Experimental studies in gastric physiology in man. II. A study of pyloric control. Role of acid and alkali. Surg. Gynecol. Obstet., 1934, 58, 935-955.

1398. Shay, H., Gershon-Cohen, R., Fels, S. S. & Siplet, H. Influence of glucose on response of stomach to test meals. Am. J. Dig. Dis., 1942, 9, 363-367.

1399. Sheenhan, D. The effect of cortical stimulation on the gastric movements of the monkey. J. Physiol., 1934, 83, 177-185.

1400. Sheiko, V. Z. Atony of the "thoracic" stomach and pyloric spasm in relation with vagotomy in Lewis′ operation. Klin. Khir., 1968, 9, 11-16.

1401. Sheiner, H. J. Gastric emptying tests in man. Gut, 1975, 16, 235-247.

1402. Sheiner, H. J., Duthie, H. L., Barber, D. C. & Howlett, P. J. Proceedings: The effect of highly selective vagotomy and truncal vagotomy with pyloroplasty on gastric emptying of solid food in man. Br. J. Surg., 1975, 62, 163-164.

1403. Shevchenko, I. A. Radiotelemetric examination of pressure in the stomach and intestines. Vrach. Delo., 1966, 9, 14-18.

1404. Shevchenko, I. A. Radiotelemetric study of the acid-forming and evacuatory functions of the stomach. Voen. Med., 1973, 7, 39-42.

1405. Shevchenko, I. A. & Dovgusha, V. V. A radiotelemetric study of the evacuatory function of the stomach following food ingestion. Vpr. Pitan., 1972, 31, 20-23.

1406. Shields, R., Griffith, G. H. & Owen, G. M. Gastric emptying in health and disease. Gastroenterology, 1968, 54, 1207-1208.

1407. Shiina, S., Hiejima, K., Hirata, Y. & Oota, Y. Trial of electrokymograph. Digest of the 6th International Conference on Medical Electronics and Biol. Eng., Tokyo, 1965.

1408. Shimamoto, K., Hikosaka, H. & Koba, T. Effects of N, N′-diphenyldicarbamate-1, 1-dimethylol cyclopentane (camalon) on the gastrointestinal motility in rabbits. Acta Sch. Med. Univ. Kioto, 1966, 39, 180-184.

1409. Shin, P. C., Kim, E. S., Kim, H. S. & Kim, Y. I. Inhibition of gastric motility in cat by acetylcholine and ganglion stimulant drugs. Arch. Int. Pharmacodyn. Ther., 1968, 171, 93-101.

1410. Shina, E. & Griffith, C. A. Selective and total vagotomy without drainage: A comparative study of gastric secretion and motility in dogs. _Ann. Surg._, 1969, _169_, 326-333.

1411. Shiratori, T. Vagotomy, lateral gastric resection and gastric motility. _Jap. J. Smooth Muscle Res._, 1974, _10_, 162-167.

1412. Shiratori, T., Kuroda, S., Sugawara, K. & Hatafuka, T. Effect of vagotomy or splanchnicectomy on the motility of the canine stomach with transection-anastomosis. _Tohoku. J. Exp. Med._, 1967, _93_, 317-329.

1413. Shiratori, T., Okabayashi, T., Harata, N. & Nagaoka, K. Intraluminal lead of the gastric action potentials in human subjects: Use of a new suction-needle electrode. _Surgery_, 1969, _66_, 483-487.

1414. Shiratori, T., Okamura, T., Kuwata, H., Misaki, M. & Shimano, Y. Effect of n-butoxybentyl-hyosyamine bromide on gastrointestinal motility. _Jap. J. Smooth Muscle Res._, 1974, _10_, 224-225.

1415. Shiratori, T., Okamura, T., Misaki, M., Nakajima, H. & Inatsugi, N. Experimental study on the effect of selective proximal vagotomy on gastrointestinal motility. _Jap. J. Smooth Muscle Res._, 1940, _10_, 223-224.

1416. Shiratori, T., Okabayashi, T., Sugawara, K. & Harata, N. Manometric and electromyographic studies on the effect of transection of the canine stomach on pyloric motor function. _Tohoku. J. Exp. Med._, 1968, _94_, 69-80.

1417. Shiratori, T. & Sekine, T. Evaluation of gastrectomy from the standpoints of movement function and acid secretion. _Surg. Ther._, Osaka, 1970, _23_, 280-289.

1418. Shiratori, T., Sugawara, O. T. & Kato, E. Intraluminal conduction of gastric action potentials with a suction needle electrode. _Digest of the Seventh International Conference on Medical and Biological Engineering_, Stockholm, 1967, 23-31.

1419. Shiratori, T., Sugawara, K. & Kuroda, S. Surgical significance of pyloroplasty with special reference to electromyographic findings. _Tohoku. J. Exp. Med._, 1965, _85_, 192-200.

1420. Shiratori, T., Sugawara, K., Okabayashi, T. & Nagaoka, K. Electromyographic study on the automatism of the stomach. _Jap. J. Smooth Muscle Res._, 1966, _2_, 105.

1421. Shiratori, T., Takakura, K., Kikuchi, S., Takase, S., Kono, H. & Sasaki, M. An electromyographic study on the gastric motility in dog. Tohoku. J. Exp. Med., 1964, 84, 228-235.

1422. Shirokova, K. I., Smakov, G. M., Chistova, S. V. & Stepenko, A. S. Motor function of the pharynx, esophagus and stomach, studied by means of balloon kymography in patients with generalized form of myasthenia before and after thymectomy. Ter. Arkh., 1974, 46, 80-87.

1423. Shuba, M. F. Electrotonus in smooth muscle. Biofizika, 1961, 6, 52-60.

1424. Shulman, L. N. The state of the motor-evacuative function of the stomach in patients with the catatonic form of schizophrenia. Zh. Neuropatol. Psikhiatr., 1966, 66, 1550-1555.

1425. Sick, K. & Tedesko, F. Studien uber Magenbewegung mit besonderer Berucksichtigung der Ausdehnungsfahigkeit des Hauptmagens (Fundus). Deutsch. Arch. f. klin. Med., 1908, 92, 416-451.

1426. Sikor, Z. G. Motor activity of a resected stomach under conditions of natural ingestion of food (based on electrogastrographic data). Klin. Med., Mosk., 1966, 44, 67-72.

1427. Sikor, Z. G. Gastric motility at the immediate postoperative period. Khirurgiia, Mosk., 1969, 45, 76-80.

1428. Silar, J., Krajsk, A. V., Bnova, M., Mottl, V., Pirk, F., Skala, I. & Skala, J. Methodological experiences in the study of gastric evacuation dynamics by means of an 113M in emitter and scintillation camera. Cas. Lek. Cesk., 1973, 112, 455-458.

1429. Silber, W. Progress in gastro-intestinal motility. S. Afr. Med. J., 1971, 45, 1144-1145.

1430. Silverstone, J. T., Smith, G. P. & Stunkard, A. J. Gastric pressures recorded by a telemetering capsule. Am. J. Dig. Dis., 1968, 13, 615-618.

1431. Simba, T. Control of the movement of the digestive tract. Jap. J. Smooth Muscle Res., 1974, 10, 134-137.

1432. Simonian, A. T. & Nanagulian, O. A. Electrogastrographic changes in peptic ulcer and gastritis (preliminary report). Zh. Eksp. Klin. Med., 1967, 7, 70-75.

1433. Singh, I. Release of hystamine on stimulation of nerves to frog's stomach muscle and its significance. Arch. Int. Pharmacodyn., 1963, 142, 522-532.

1434. Singh, I. Release of substance P on nervous and electrical stimulation of frog's stomach muscle. Arch. Int. Pharmacodyn., 1963, 143, 138-143.

1435. Singh, I. Responses of frog's stomach muscle produced by nervous stimulation presumably due to release of 5-hydroxytryptamine. Arch. Int. Pharmacodyn., 1963, 141, 67-74.

1436. Singh, I. Seasonal variations in the nature of neurotransmitters in a frog vagus-stomach muscle preparation. Arch. Int. Physiol., 1964, 72, 843-851.

1437. Skala, I. & Pirk, F. Roentgenological study of gastrointestinal motility following administration of medium chain triglycerides in comparison with olive oil. Cesk. Gastroenterol. Vyz., 1968, 22, 174-179.

1438. Skoryna, S. C. Pathophysiology of Peptic Ulcers. Philadephia: Lippincott, 1964.

1439. Slaughter, D., Goddard, A. B. & Henderson, W. M. Some new aspects of morphine action: Effect on stomach. J. Pharmacol., 1942, 76, 301-308.

1440. Sleeth, C. K. & Van Liere, E. J. The effect of environmental temperature on the emptying time of the stomach. Am. J. Physiol., 1937, 118, 272-275.

1441. Slutzky, B. & Anderson, A. C. Digestive system. Ann. Rev. Physiol., 1944, 6, 225-246.

1442. Smejkal, V. Effect of reserpine on some functions of the digestive tract in rats-- on gastric secretion and gastrointestinal motility. Cesk. Gastroenterol. Vyz., 1967, 21, 209-215.

1443. Smith, A. N. & Dogg, D. Effect of gastrin II on the motility of the gastrointestinal tract. Lancet, 1966, 1, 403-404.

1444. Smith, A. N. & Giannakos, V. Telemetering in the gastrointestinal tract. Surg. Annu., 1971, 3, 163-180.

1445. Smith, A. N., Kirwan, W. O. & Shariff, S. Motility effects of operations performed for diverticular disease. Proc. Royal Soc. Med., 1974, 67, 1041-1043.

1446. Smith, A. N. & Ridgeway, M. Use of telemetering capsules in disorders of alimentary tract. I. Technique for recording pressure changes. Gut, 1962, 3, 366-371. II. Gut, 1962, 3, 371-376.

1447. Smith, A. N. & Ridgeway, M. Recording pressures in the gastrointestinal tract using a telemetering capsule. Soc. Int. Chir., 1963, 1, 97.

1448. Smith, A. W. M. & Code, C. F. Gastric antral motility after ingestion of food. Am. J. Physiol., 1956, 187, 633.

1449. Smith, A. W. M. & Code, C. F. The effect of an ordinary and of an excessively fatty breakfast on human gastric antral motility. Gastroenterology, 1958, 35, 398-405.

1450. Smith, A. W. M., Code, C. F. & Schlegel, J. F. Simultaneous cineradiographic and kymographic studies of human gastric antral motility. J. Appl. Physiol., 1957, 11, 12-16.

1451. Smith, H. W. Intraluminal pressures from upper gastrointestinal tract. II. Correlations with gastroduodenal motor activity in normal subjects and patients with ulcer distress. Gastroenterology, 1957, 32, 1025-1047.

1452. Smith, H. W. & Texter, E. C., Jr. Characteristics of phasic intraluminal pressure waves of stomach and duodenum. Am. J. Dig. Dis., 1957, 2, 318-326.

1453. Smith, J. M. Changes in gastrointestinal motility during conditioned supression. Proceedings of the 80th Annual Convention of the American Psychological Association, 1972, 7, 650-652.

1454. Smith, J. M., Renault, P. F. & Schuster, C. R. Operant schedule control of differential changes in gastrointestinal motility. Proceedings of the 81st Annual Convention of the American Psychological Association, 1973, 81, 849-850.

1455. Snowdon, C. T. Motivation and the control of meal parameters with oral and intragastric feeding. J. Comp. Physiol. Psychol., 1969, 69, 91-100.

1456. Snowdon, C. T. Gastrointestinal sensory and motor control of food intake. J. Comp. Physiol. Psychol., 1970, 71, 68-76.

1457. Sobakin, M. A. & Mishin, L. N. Novyye dannyye metodiki i tekhaiki elektrogastrografii. (New data, methods and techniques in électrogastrography.) In: Materialy po obmenu opytom i nauchymi dostizheniyami v meditsinskoy promyshlennosti. (Material toward exchange of experience and scientific achievement in the medical industry.) Moscow, 1958.

1458. Sobakin, M. A., Smirnov, I. P. & Mishin, L. N. Electrogastrography. IRE Trans. Biomed. Electr., 1962, BME-9, 129-132.

Surface recording of stomach potentials requires high amplification along with attenuation of very slow and very high frequency potentials. Using such a device records have been taken from 61 healthy subjects and 164 patients suffering gastric disorder. Three/minute electrical rhythms were observed in thin and obese individuals. Ulcer caused no change in the 3/min rhythm but pyloric stenosis produced a doubling of the amplitude. Cancer of the stomach caused a breakdown of the basic rhythm.

1459. Soergel, K. H. Effect of gastrointestinal hormone on gastrointestinal motility and absorption. Verh. Dtsch. Ges. Inn. Med., 1974, 80, 385-392.

1460. Sokolovskii, V. D., Poliakov, N. G. & Nazarevich, A. G. Acute atony of the stomach and urinary bladder in diabetic coma. Probl. Endokrinol. Gormonoter, 1966, 12, 57-58.

1461. Sorokin, G. A. The motor and secretory function of the stomach in toxic goiter patients before and after surgery. Khirurgiia, 1969, 45, 36-39.

1462. Sperelakis, N. Contraction of depolarized smooth muscle by electric fields. Am. J. Physiol., 1962, 202, 731-742.

1463. Sperelakis, N. & Tarr, M. Weak electronic interaction between neighboring visceral smooth muscle cells. Am. J. Physiol., 1965, 208, 737-747.

1464. Sporn, J. & Nicheles, H. Effect of glucagon on gastrointestinal motility. Am. J. Physiol., 1956, 187, 634.

1465. Springfield, A. C., Weddle, C. O., II, Ormsbee, H. S., III, Barreras, R. F. & Bass, P. Influence of truncal vagotomy on canine gastric emptying of liquids. Am. J. Surg., 1974, 128, 678-682.

1466. Stadaas, J. Intragastric pressure/volume relationship before and after proximal gastric vagotomy. Scand. J. Gastroenterol., 1975, 10, 129-134.

1467. Stadaas, J. Intragastric pressure/volume relationship in the normal human stomach. Scand. J. Gastroenterol., 1975, 10, 135-140.

1468. Stadaas, J. & Aune, S. Intagastric pressure/volume relationship before and after vagotomy. Acta Chir. Scand., 1970, 136, 611.

1469. Stadaas, J. & Aune, S. Motility of the stomach. Nord. Med., 1971, 85, 601.

1470. Stadaas, J., Aune, S. & Haffner, J. F. Effects of proximal gastric vagotomy on intragastric pressure and adaptation in pigs. Scand. J. Gastroenterol., 1974, 9, 479-485.

1471. Stadaas, J. & Schrumpf, E. Intragastric pressure-volume relationship and gastrin release after distention of separate parts of the stomach. Scand. J. Gastroenterol., 1974, 9, 781-785.

1472. Stadaas, J., Schrumpf, E. & Haffner, J. F. The effect of gastric distention on gastric motility and serum gastrin in acutely vagotomized pigs. Scand. J. Gastroenterol., 1974, 9, 127-131.

1473. Stanciv, C. & Bennett, J. R. Effects of pancreozymin and glucagon or gastro-duodenal motility in normal subjects. Rev. Med. Chir. Soc. Med. Nat. Iasi., 1974, 78, 609-613.

1474. Stanciv, C. & Bennett, J. R. The general pattern of gastro-duodenal motility: 24 hour recordings in normal subjects. Rev. Med. Chir. Soc. Med. Nat. Iasi., 1975, 79, 31-36.

1475. Stavney, L. S., Kato, T., Griffith, C. A., Nyhus, L. M. & Harkins, H. N. A physiologie study of motility changes following selective gastric vagotomy. J. Surg. Res., 1963, 3, 390-394.

1476. Steggerda, F. R. The digestive system. Ann. Rev. Physiol., 1955, 17, 129-154.

1477. Steggerda, F. R. & Clark, W. C. Pressure and activity recordings along the gastrointestinal tract in man. Am. J. Physiol., 1948, 155, 470-471.

1478. Steggerda, F. R. & Clark, W. C. Pressure and activity recordings in the pylorus. Fed. Proc., 1949, 8, 151.

1479. Steigmann, F., Kaminski, L. & Dolehide, R. H. Effects of smoking tobacco on gastric acidity and motility of hospital controls and patients with peptic ulcer. Am. J. Gastroenterol., 1954, 22, 399-409.

1480. Stein, J. F., Jr., Meyer, K. A. & Steigmann, F. Studies of vagotomy in treatment of peptic ulcer; changes in gastric motility and the effect of drugs on motility following complete vagotomy. Surg. Gynecol. Obstet., 1948, 87, 465-471.

1481. Stemper, T. J. Gastric emptying and its relationship to antral contractile activity. Gastroenterology, 1975, 69, 649-653.

1482. Stephens, J. R., Woolson, R. F. & Cooke, A. R. Effects of essential and nonessential amino acids on gastric emptying in the dog. Gastroenterology, 1975, 69, 920-927.

1483. Stern, R. M. Effects of variation in visual and auditory stimulation on gastrointestinal motility. Psych. Rep., 1964, 14, 799-802.

The purpose of this experiment was to investigate the effects of three different stimulus conditions on gastrointestinal motility as measured by electrodes on the surface of the abdomen. The 16 subjects were each run for two 40-min. sessions, during which they were instructed to lie quietly on a cot. Amplitude differentiated between a group that received diffuse stimulation and groups that received normal stimulation and no stimulation, with the former showing greater activity throughout the session. The increase in gastrointestinal activity for subjects in the diffuse stimulus group is attributed to the novelty and ambiguity of the stimulus situation.

1484. Stern, R. M. Effects of contrast in stimulation on gastrointestinal motility. Psych. Rep., 1965, 16, 156-158.

The purpose of this study was to determine the effects of contrast in stimulation, or in other words, the effects of the immediately preceding level of stimulation on the gastrointestinal response to a moderate level of stimulation. As was hypothesized, a change to a moderate level of stimulation resulted in a decrease in g.i. motility for a group of 20 male subjects who were responding at a high level of high stimulation, and an increase in level of responding for another group of 20 male subjects who were responding at a previously low level to low stimulation. For various autonomic measures there is an immediate reversal in relative level of responding with change in level of stimulation, whereas for g.i. motility there is a 2- to 3-min. increase in activity for both groups following any change in stimulation and then a reversal in level of g.i. responding.

1485. Stern, R. M. A re-examination of the effects of response-contingent aversive tones on gastrointestinal activity. Psychophysiology, 1966, 2, 217-223.

The purpose of this study was to compare the response contingent and random aversive auditory stimulation on gastrointestinal motility. subjects were presented with four consecutive 10-min treatment periods in the following order: response-contingent (R-C), random, R-C random. During R-C, S could avoid the 5-sec 90 db tone, which was programmed to come on every 30 sec, by pressing a key once during the 5 sec preceding the tone. During the random period, S had no control over the presentation of the tone. The results indicated that during R-C, amplitude of gastrointestinal motility was significantly greater and latency briefer than under random conditions. These results are accounted for in terms of the dual role of the tone in the R-C condition, i.e. it is aversive and it signifies to S that his time estimation was incorrect.

1486. Stevens, J. K. & Worrall, N. External recording of gastric activity: The electrogastrogram. Physiol. Psychol., 1974, 2, 175-180.

Reports in the American and Russian literature have suggested that recording of electrical activity from electrodes on the surface of the abdomen (electrogastrogram, EGG) might be used as a means of measuring gastric motility for psychophysiological research and for clinical purposes. However, the complexity of the recordings has presented problems in specifying relations between the EGG and other measures of gastric activity. In this study, cats are used to demonstrate the validity of the EGG as a measure of gastric activity by taking recordings from a strain guage on the stomach wall simultaneously with electrical recordings from the surface of the abdomen. The present report also demonstrates that autocorrelation and Fourier analysis may be used to reduce record complexity and to provide information unattainable by conventional methods.

1487. Stevens, N. C. Auscultation of abdomen: Aid to diagnosis. New Engl. J. Med., 1934, 211, 108.

1488. Stevens, N. C. Auscultation of abdomen. New Engl. J. Med., 1936, 215, 22.

1489. Stickney, J. C. & Northrup, D. W. Effect of gastric emptying upon propulsive motility of small intestine of rats. Proc. Soc. Exp. Biol. Med., 1959, 101, 582-583.

1490. Stickney, J. C., Northrup, D. W. & Van Liere, E. J. Effect of increase CO2 in inspired air on gastric emptying in dogs. Am. J. Physiol., 1948, 155, 471.

1491. Stoddard, C. J., Brown, B. H. & Duthie, H. L. Proceedings: The effects of varying the extent of vagotomy on the canine gastric and duodenal myoelectrical activity. Br. J. Surg., 1973, 60, 913.

1492. Stoddard, C. J. & Duthie, H. L. Proceedings: The changes in gastroduodenal myoelectrical activity after varying degrees of vagal denervation. Gut, 1973, 14, 824.

1493. Stoddard, C. J. & Duthie, H. L. Long term effects of vagotomy on gastric myoelectrical activity. Br. J. Surg., 1974, 61, 912.

1494. Stoddard, C. J. & Duthie, H. L. Proceedings: Effects of varying degrees of vagotomy on the gastroduodenal response to food. Gut, 1974, 15, 348.

1495. Stoddard, C. J., Smallwood, R., Brown, B. H. & Duthie, H. L. The immediate and delayed effects of different types of vagotomy on human gastric myoelectrical activity. Gut, 1975, 16, 167-170.

1496. Stordy, S. N., Greig, J. H. & Bogoch, A. The steak and barium meal: A method for evaluating gastric emptying after partial gastrectomy. Am. J. Dig. Dis., 1969, 14, 463-469.

1497. Strom, G. & Uvnas, B. Motor responses of gastrointestinal tract and bladder to topical stimulation of the frontal lobe, basal ganglia and hypothalamus in the cat. Acta Physiol. Scand., 1950, 21, 90-104.

1498. Struchkov, V. I. & Sundushnikov, V. N. Electrogastrography in gastrointestinal hemorrhage. Khirurgiia, Mosk., 1971, 47, 113-117.

As the result of studying the electrogastrograms

in 87 patients with gastrointestinal hemorrhages the authors established that the character of the curve depends on the pathological process which caused the bleeding. In hemorrhages of ulcerous etiology there is a hyperkinetic type of curve with sharply increased voltage in a preserved rhythm of contraction. Electrogastrographic investigation together with other methods is conducive to improving the diagnosis of gastrointestinal hemorrhages.

1499. Stubel, H. Der Erregungsvorgang in der magenmuskulatur, nach versuchen am frosch und am Vogelmagen. Arch. f. d. ges. physiol., 1912, 143, 381-394.

1500. Stunkard, A. J. Hunger and satiety. Am. J. Psychiat., 1961, 118, 212-217.

1501. Stunkard, A. J. & Fox, S. The relationship of gastric motility and hunger. A summary of the evidence. Psychosom. Med., 1971, 33, 123-134.

1502. Stunkard, A. J. & Koch, C. R. The interpretation of gastric motility. I. Apparent bias in the reports of hunger by obese persons. Arch. Gen. Psychiat., 1964, 11, 74-82.

1503. Stunkard, A. J. & Koch, C. R. The interpretation of gastric motility. II. Sensitivity and bias in the perception of gasrtic motility. Arch. Gen. Psychiat., 1964, 11, 82.

1504. Stunkard, A. J. & Reinhard, A. Effects of an intragastric balloon on gastric pressures in man. J. Appl. Physiol., 1965, 20, 1345.

1505. Stunkard, A. J., Van Itallie, T. B. & Reiss, B. B. Mechanism of satiety: Effect of glucagon on gastric hunger contractions in man. Proc. Soc. Exp. Biol. Med., 1955, 89, 258-261.

1506. Stunkard, A. J. & Wolff, H. G. Correlation of arteriovenous glucose difference, gastric hunger contraction and the experience of hunger in man. Fed. Proc., 1954, 13, 147.

1507. Sturgeon, A. M., Henschel, A. & Keys, A. The emptying of the human stomach as influenced by the nutrient composition of the meal. Fed. Proc., 1945, 4, 68-69.

1508. Sudakov, K. V. & Rogacheva, S. K. The afferent and efferent activity of the fibers of the vagus nerve during fasting and after taking food. Fed. Proc., 1963, 22, Trans. Suppl. T306-310.

1509. Sudsaneth, S. & Mayer, J. Relation of metabolic events to gastric contractions in the rat. Am. J. Physiol., 1959, 197, 269-273.

1510. Suenaga, K. & Nishi, S. Study of the stomach muscle by microelectrode technique. J. Jap. Physiol. Soc., 1956, 20, 269-270.

1511. Sugano, O. Studies on the action potential of the gastric wall--with special reference to observations of the anastomotic area. Acta Med., (Fukuoka), 1967, 37, 445-459.

1512. Sugawara, I., Isaza, J., Curt, J. & Woodward, E. R. Effect of secretin and cholecystokinin on gastric motility. Am. J. Physiol., 1969, 217, 1633.

1513. Sugawara, K. An electromyographic study on the motility of canine stomach after transection and end-to-end anastomosis. Tohoku. J. Exp. Med., 1964, 84, 113-124.

1514. Sugawara, K., Chawla, R. C. & Eisenberg, M. M. Simultaneous observation of gastric motor and secretory response to vagal stimulation in normal and diabetic dogs. Surgery, 1977, 70, 325-333.

1515. Sugawara, K., Isaza, J. & Woodward, E. R. Effect of gastrin on gastric motor activity. Gastroenterology, 1969, 57, 649.

1516. Sugawara, K., Isaza, J., Woodward, E. R. & Drugstedt, L. K. The short term effect of vagotomy on gastric motility. Arch. Surg., 1969, 99, 1.

1517. Sugawara, K., Kato, M., Yanagisawa, T., Kawai, K. & Fukumoto, K. Effect of stomach dilatation on rhythmic stomach contraction. Jap. J. Smooth Muscle Res., 1974, 10, 226-227.

1518. Sukhodolo, V. D., Didovenko, A. O., Gainutdinova, R. K. & Popova, L. G. Appearance of periodic rhythm components in the activity of the salivary glands, stomach and duodenum in hunger and after intake of small quantities of food. Fiziol. Zh. S.S.S.R., 1970, 56, 1694-1702.

1519. Suyama, T. & Kawamura, I. Secretin and gastrointestinal motility. Clin. Endocrinol., Tokyo, 1975, 23, 747-751.

1520. Suzuki, K. Electromyographic study of motility of the human digestive tract after various abdominal operations. Jap. J. Smooth Muscle Res., 1968, 4, 89-102.

1521. Swan, K. G., Knoturek, S. J., Jacobson, E. D. & Grossman, M. I. Inhibition of gastric secretion and motility by fat in the intestine. Proc. Soc. Exp. Biol. Med., 1966, 121, 840-844.

1522. Swift, M. N., Taketa, I. T. & Bond, V. P. Delayed gastric emptying in rats after whole and partial body x-irradiation. Am. J. Physiol., 1955, 182, 479-486.

1523. Synovets, A. S. & Sorokin, G. A. Electrogastrographic method of the stomach motility investigation in patients with toxic goiter. Vestn. Khir., 1969, 103, 31-33.

The authors have studied the functional state of the stomach in 117 patients with thyrotoxicosis prior to subtotal thyroidectomy and in 95 of these patients after the operation. Electrogastrographic method of investigation of the gastric motor function was used as the most precise and objective one. It was found that in patients with thyrotoxicosis grave but reversible motor disturbances could occur. It necessitates their recognition in due course and appropriate treatment.

1524. Szekely, A., Major, T. & Romvari, H. Effect of pentagastrin on gastric motility (radiokymographic studies). Fortschr. Geb. Roentgenstr. Nuklearmed., 1969, 111, 841-846.

1525. Szurszewski, J. H. Mechanism of action of pentagastrin and acetylcholine on the longitudinal muscle of the canine antrum. J. Physiol., 1975, 252, 335-361.

1526. Takeuchi, M., Aizawa, I. & Tabe, M. Study of gastrointestinal motility using an extraluminal force transducer. III. Gastric movement in a conscious state. Jap. J. Smooth Muscle Res., 1974, 10, 234-235.

1527. Tamarit-Torres, J. Una nueva ley de evacuacion gastrica. Arch. Med. Exptl., 1955, 18, 307-328.

1528. Tamarit-Torres, J. & Salamanca, F. E., Jr. El funcionamiento gastrico en la ulcera gastro-duodenal. Trabajos Inst. Nacional Cienc. Med., 1950, 13, 89-113.

1529. Tamarit-Torres, J., Salamanca, F. E., Jr. & Castro-Rial, M. El ritmo de apertura pilorica en el ulcus gastroduodenal. Arch. Med. Exptl., Madrid, 1954, 12, 99-118.

1530. Tanaka, M. D. Y. Clinical application of the intracanal gastro-electromyography. Digest of the 6th International Conference on Medical Electronics and Biological Engineering, Tokyo, 1965.

1531. Tanaka, N. The vagus nerve and gastric motility. _Jap. J. Smooth Muscle Res._, 1974, _10_, 157-159.

1532. Tanaka, Y. & Oi, M. Recording of gastric electrical potential with a suction electrode. _Proc. 6th Intern. Conf. Med. Biol. Eng._, 1965, 482.

1533. Tanaka, Y. & Yanigasawa, K. The electromyography for the stomach and its clinical application. _Jap. J. Electron. Biol. Eng._, 1963, _1_, 309.

1534. Tanaka, Y., Yanigasawa, M. & Oi, M. Recording of gastric electrical potential with a suction electrode. _Proc. 7th Intern. Conf. Med. Biol. Eng._, 1965.

1535. Tauber, O., Schmid, E., Heinkel, K. & Henning, N. Effects of antiinflammatory drugs on the secretion and motility of the human stomach. _Arzneim. Forsch._, 1966, _16_, 155-162.

1536. Temiaeva, M. B. Restitutsiia sekretsii; dvizhemia zheludka v usloviiakh regeneratsii bluzhdaiushchikh nervov u sobaki. (Restitution of secretion and movement of stomach under conditions of regeneration of vagus nerve.) _Fiziol. Zh. S.S.S.R._, 1947, _33_, 611-626.

1537. Templeton, R. D. A new method of graphic representation of motor activity of the stomach in intact animals. _Am. J. Physiol._, 1928, _85_, 512-517.

1538. Templeton, R. D. & Johnson, V. Further observations on the nature of hunger contractions in man. _Am. J. Physiol._, 1929, _88_, 173.

1539. Test, C. E., Nichols, R. L., Landau, R. L., Ricketts, H. T. & Loughead, R. Study of "flat" oral glucose tolerance curve and its relation ro rate of emptying of stomach. _Am. J. Med. Sci._, 1956, _231_, 69-74.

1540. Texter, E. C., Jr. Motility in the gastrointestinal tract. _J. Am. Med. Assoc._, 1963, _184_, 640-647.

1541. Texter, E. C., Jr. The control of gastrointestinal motor activity. _Am. J. Dig. Dis._, 1964, _9_, 585-598.

1542. Texter, E. C., Jr. Motility comes of age. _Am. J. Dig. Dis._, 1971, _16_, 681-688.

1543. Texter, E. C., Jr., Vantrappen, G., Liemer, M. D. & Barborka, C. J. Methods for studying gastrointestinal motility. _Quart. Bull. Northwestern Univ. Med. School_, 1958, _32_, 281-290.

1544. Theile, P. Zur Radiologie des Sauglingsmagens mit besonderer Berucksichtigung der Gestalts- und Lageveranderungen dieses Organes bei der Fullung und Entleerung. Ztschr. f. Kinderh., 1917, 15, 152-212.

1545. Thomas, J. E. The mechanism of gastric evacuation. J. Am. Med. Assoc., 1931, 97, 1663-1668.

1546. Thomas, J. E. Gastric inhibition caused by amino acids in the small intestine. Am. J. Physiol., 1942, 135, 609-613.

1547. Thomas, J. E. Digestion and the nervous system; a review of the literature. Am. J. Dig. Dis., 1943, 10, 201-206.

1548. Thomas, J. E. The intestinal pH threshold for regulation of gastric emptying. Fed. Proc., 1947, 6, 214.

1549. Thomas, J. E. Recent advances in gastrointestinal physiology. Gastroenterology, 1949, 12, 545-560.

1550. Thomas, J. E. The autonomic nervous system in gastrointestinal disease. J. Am. Med. Assoc., 1955, 157, 209-212.

1551. Thomas, J. E. Mechanics and regulation of gastric emptying. Physiol. Rev., 1957, 37, 453-474.

1552. Thomas, J. E. & Baldwin, M. V. Pathways and mechanisms of regulation of gastric motility. In C. F. Code (Ed.), Handbook of Physiology. Sect. 6, Alimentary Canal. Vol. IV. Motility. Washington, D.C.: American Physiological Society, 1968.

1553. Thomas, J. E. & Crider, J. O. Rhythmic change in duodenal motility associated with gastric peristalsis. Am. J. Physiol., 1935, 11, 124-129.

1554. Thomas, J. E. & Crider, J. O. Inhibition of gastric peristalsis due to a substance in commercial peptones. Proc. Soc. Exp. Biol. Med., 1935, 34, 825-827.

1555. Thomas, J. E. & Crider, J. O. Rhythmic changes in duodenal motility associated with gastric peristalsis. Am. J. Physiol., 1935, 111, 124-129.

1556. Thomas, J. E. & Crider, J. O. Inhibition of gastric peristalsis due to a substance found in commercial peptones. Proc. Soc. Exp. Biol. Med., 1936, 34, 825-827.

1557. Thomas, J. E. & Crider, J. O. The regulation of gastric emptying. Virginia M. Monthly, 1937, 64, 181-185.

1558. Thomas, J. E. & Crider, J. O. Inhibition of gastric motility associated with the presence of products of protein hydrolysis in the upper small intestine. Am. J. Physiol., 1939, 126, 28-38.

1559. Thomas, J. E., Crider, J. O. & Morgan, C. J. A study of reflexes involving the pyloric sphincter and antrum and their role in gastric evacuation. Am. J. Physiol., 1934, 108, 683-700.

1560. Thomas, J. E. & Friedman, M. H. T. Digestive system. Ann. Rev. Physiol., 1949, 11, 103-138.

1561. Thomas, J. E. & Kuntz, A. A study of gastro-intestinal motility in relation to the enteric nervous system. Am. J. Physiol., 1926, 76, 606-626.

1562. Thompson, C. M. Stomach, New York: Grune & Stratton, 1967.

1563. Thouvenot, J., Harichaux, P. & Moline, J. La part des mecanismes neuro-musculaires lissess intrinsiques dans le fontionnement antropylorique. Rev. Med. Tours, 1965, 6, 263-267.

1564. Thouvenot, J. & Martin, A. L'Electrosplanchnographie. Bases theoriques. Etat actuel des vecherches. Prospectives d'avenir en pathologie digestive. Presse Med., 1969, 77, 1567-1569.

ESG opens a multitude of possibilities in digestive pathology--medical, surgical, and psychosomatic. This functional study can be carried out in a completely innocuous way, and studies can last for several days. The technique of collecting data is very direct. It consists of recording differences in electrical potential from the surface of the abdominal skin, displaying the activity of the underlying viscera. This is a type of interdisciplinary research. The progress thus far merits the establishment of an appropriate research facility to extend the research.

1565. Thouvenot, J., Martin, A. & Gaudeau, C. Reconnaissance des processus dynamiques dans l'electrosplanchnogramme, ESG. Gen. Biol. Med., 1970, 1, 56-59.

1566. Thouvenot, J., Martin, A. & Touron, P. Derived potentials in the abdominal wall; relation with digestive visceral motility in humans. J. Physiol., 1968, 60, 556-557.

1567. Thouvenot, J. & Penaud, J. L'Electrosplanchnographie. Paris: Sarget Laboratoires, 1975.

The history of surface recording of abdominal potentials is reviewed. The technique used is presented along with descriptions of multiple electrode locations. The electrical records are compared to mechanographs in the rate and characteristic patterns of electrical activity are described for various locations in humans. Records are presented showing characterstic stomach waves and faster duodenal waves.

1568. Thouvenot, J., Penaud, J., Arhan, P. & Faverdin, C. Problems raised by electric activity analysis of digestive viscera collected by cutaneous derivation. J. Physiol., (Paris), 1971, 63, 102A-103A.

This study is related to those that have been carried out on humans. It is concerned with how to use surface electrophysiological recording to study the internal source and the transmission of the activity recorded at the skin surface. In conclusion, the criterion "frequency" appears to be the most suitable for analysis of the complex activities that it is possible to record at the surface of the skin.

1569. Thouvenot, J., Tonkovic, S. & Penaud, J. Electrosplanchnography--Method for the electrophysiological exploration of the digestive tract. Acta Med. Iug., 1973, 27, 227-247.

A method for the electrophysiological exploration of the digestive tract by means of cutaneous electrodes is described. We have named it Electrosplanchnography (E.S.G.). The technique of the signals acquisition and the analysis of recorded signals is presented. The band of frequencies, the most fundamental in E.S.G., is inferior to 1 c/s. This implies some problems in the data acquisition such as the augmentation of the electrode resistance due to electrode polarization, the use of the amplifiers with a high time constants (in order of 5-10 seconds), the influence of the psychogalvanic reflex, etc. We have utilised a comparison technique in simultaneously recording the electrical activity "in situ" and on the skin surface to evidence typical waveforms of different digestive organs. The results demonstrate that there exist a bond between mechanical and electrical activity but that a reciprocal relation does not necessarily exist. This constitute one of the most difficult problems in the analysis of traces. However, even in the absence of a visual recognition of typical waveforms, the numerical analysis of tracing, after digitalization, most often allows their identification.

1570. Tidwell, H. C. & Cameron, E. S. Relation between chemical structure of fats and their ability to produce gastric inhibition. Bull. Johns Hopkins Hosp., 1942, 70, 362-369.

1571. Tikhonov, M. K. Diagnostic value of electrogastrography in diseases of the stomach. Voen. Med., 1970, 4, 50-52.

Concerned with chronic gastritis and functional disturbances of the stomach. For the study of the functional state of the stomach, the character of its secretory and motor activity has great significance. Author studied patients with chronic gastritis and functional disturbances. Used EGS-3 30 minutes after experimental meal. Active electrode on anterior abdominal wall in region of pylorus. Patients showed hyper-motor activity.

1572. Titchen, D. A. Reflex stimulation and inhibition of reticulum contractions in the ruminant stomach. J. Physiol., 1958, 141, 1-21.

1573. Titchen, D. A. The production of rumen and reticulum contractions in decerebrate preparations of sheep and goat. J. Physiol., 1960, 151, 139-153.

1574. Toda, N. & Tobe, T. Studies on stereoroentgenography of motile organs. II. Gastrointestinal tract. Nippon Acta Radiol., 1967, 27, 391-405.

1575. Todd, T. W. Behavior Patterns of the Alimentary Tract. The Beamont Foundation Lectures. Series no. 9, Baltimore: Williams & Wilkins, 1930.

1576. Todd, T. W. Practical control of G I motility. Proc. Intern. Assemb. Inter State-Post Grad. Med. Assoc. N. Amer., (1930), 1931, 6, 549-554.

1577. Tomita, T. Electrical properties of mammalian smooth muscle. In E. Bulbring, A. F. Brading, A. W. Jones, & T. Tomita (Eds.), Smooth Muscle. London: Arnold, 1971.

1578. Tomov, T. & Keling, K. On the motility of the rumen and belching in small ruminants. Dokl. Bolg. Akad. Navk., 1965, 18, 971-974.

1579. Tonkovic, S., Penaud, J., Thouvenot, J. & Mountafian, J. P. Une application de l'analyse spectrale dans le traitment des signaux electrosplanchnographiques. Med. Biol. Eng., 1975, 266-271.

In this paper we have tested an application of spectral analysis to the problem of the processing of bioelectric visceral signals. As has been shown, bioelectric visceral signals (electrosplanchnogramm--e.s.g.) derived by means of cutaneous electrodes are wide-sense stationary in an 'defined interval' and so, by use of spectral analysis, the characteristic rhythms of the different abdominal organs can be found. Eight different subjects are studied and a very good agreement between the frequencies found using spectral analysis and the proper rhythms of the diffferent organs otherwise obtained (measurements in situ with intra-intestinal probes etc.) is shown. An analysis of the power spectra shows that it is possible to determine typical frequency bands for the different organs' activities. This allows us, using suitable filtering, to identify their activities, even in the absence of visual recognition of typical waveforms. A method for locating the electrodes is also discussed.

1580. Torek, P. Abdominal ausculation. Am. Pract., 1947, 1, 375.

1581. Torgersen, J. The muscular build and movements of the stomach and duodenal bulb. Acta Radiol., 1942, 45, 1-191.

1582. Towbin, E. J. Gastric distention as a factor in the satiation of thirst. Fed. Proc., 1949, 8, 158.

1583. Toyama, T. & Blickman, J. R. Serial roentgenographic observations on the passage of foods through the human stomach. Radiol. Clin. Biol., 1971, 40, 29-38.

1584. Toyama, T., Tanaka, H. & Akiba, H. Chronological tracing of the movement of food particles in the human stomach--In a normal person on a regular diet. Rinsho Hoshasen, 1968, 13, 524-530.

1585. Transy, M. F., Benhayem, S., Probst, S. & Jordan, J. S. The effects of methyl methacrylate vapor on gastric motor function. J. Am. Dent. Assoc., 1974, 89, 372-376.

1586. Tristan, T. A. Prospectus of radiologic techniques for gastroenterology. Am. J. Dig. Dis., 1963, 8, 33-52.

1587. Tsang, Y. C. Hunger motivation in gastrectomized rats. J. Comp. Psychol., 1938, 26, 1-17.

1588. Tsellarius, A. N., Sorokin, G. A., Strizhevskii, A. N. & Liubarskii, R. I. The value of examining the motor-evacuation activity of the stomach for assessing the status of patients with toxic goiter during the preoperative period. Klin. Khir., 1967, 4, 30-34.

1589. Tsukamoto, M., Herrera, F., Kenys, D. R., Emas, G. S., Woodward, E. R. & Eisenberg, M. M. Effect of vagal stimulation by 2-deoxy-d-glucose and insulin on gastric motility in dogs. Am. Surg., 1967, 165, 605-608.

1590. Tumpeer, K. H. & Phillips, B. Hyperperistaltic electrographic effects. Am. J. Med. Sci., 1932, 184, 831-836.

1591. Turaev, R. N. & Berdov, B. A. Methods of radioisotope scanning for study of the motor-evacuatory function of the stump or a resected stomach. Eksp. Khir. Anesteziol., 1971, 16, 38-39.

1592. Tuttle, S. G., Bettarello, A. & Grossman, M. I. Esophageal acid perfusion test and a gastroesophageal reflux test in patients with esophagitis. Gastroenterology, 1960, 38, 861-872.

1593. Tuzhilin, S. A. & Lukash, L. K. Motor-evacuatory function of gastro-intestinal tract in chronic pancreatitis. Vrach. Delo., 1967, 10, 51-54.

1594. Ustach, T. J., Tobon, F., Hambrecht, T., Bass, D. D. & Schuster, M. M. Electrophysiological aspects of human sphincter function. J. Clin. Invest., 1970, 49, 41.

1595. Ustinov, V. N. Configuration of gastric and duodenal smooth muscle biopotentials. Fiziol. Zh. S.S.S.R., 1974, 60, 961-971.

1596. Ustinov, V. N. & Kotelnikova, V. I. Interrelation of the bioelectric and motor activity of the stomach, duodenum and gallbladder. Fiziol. Zh. S.S.S.R., 1971, 57, 284-291.

1597. Ustinov, V. N. & Papazova, M. Method of bioelectric stimulation of gastric smooth muscles. Fiziol. Zh. S.S.S.R., 1974, 60, 831-836.

1598. Uvnas, B. Digestion. Ann. Rev. Physiol., 1956, 18, 145-158.

1599. Vagne, M. & Andre, C. The effect of secretin on gastric emptying in man. Gastroenterology, 1971, 60, 421-424.

1600. Vainshtein, S. G. The effect of mineralocorticoids on the motor and secretory function of the stomach. Patol. Fiziol. Eksp. Ter., 1966, 10, 79-82.

1601. Vakaliuk, P. M. & Makos, R. P. Indices of electrogastrography and uropepsin level in patients with myocardial infarct. Vrach. Delo., 1972, 5, 66-68.

The electrogastrogram of patients suffering from myocardial infarction with the abdominal syndrome showed characteristic changes manifested in "volleys" of frequent but arrhythmic waves with an amplitude 3-4 times above normal. Uropepsin levels were twice higher than in healthy persons. Following repeated administration of morphine, diaphhyllin, nitrites, phentanil with properidol there occured a sharp inhibition of the electric potentials and uropepsin. The above changes were observed in 33 patients with pain syndrome prevailing in the clinical picture and they tended to normalize on the 33-40 days. In 26 other patients with acute cardiac and vascular insufficiency, on admission and during the first 5-9 days the electrogastrograms were characterized by hypokinetic dyskinesia, and by a low uropepsin level. Here normalization occurred later.

1602. Vane, J. R. The relative activities of some tryptamine analogues on the isolated rat stomach preparation. Br. J. Pharmacol., 1959, 14, 87-98.

1603. Van Liere, E. J. & Northrup, D. W. The emptying time of the stomach of old people. Am. J. Physiol., 1941, 134, 719-722.

1604. Van Liere, E. J. & Northrup, D. W. Effect of water taken with meals on gastric emptying. Gastroenterology, 1944, 2, 195-200.

1605. Van Liere, E. J. & Northrup, D. W. Effect of nalorphine on gastric emptying in man. Proc. Soc. Exp. Biol. Med., 1956, 91, 619-620.

1606. Van Liere, E. J., Northrup, D. W. & Canada, W. H. Administration of carbonesin and gastric emptying in man. J. Appl. Physiol., 1955, 8, 119-120.

1607. Van Liere, E. J., Northrup, D. W. & Stickney, J. C. Effect of glucose on motility of stomach and small intestine. Gastroenterology, 1946, 7, 218-223.

1608. Van Liere, E. J., Sleeth, C. K. & Northrup, D. W. The relation of the size of the meal to the emptying time of the human stomach. Am. J. Physiol., 1937, 119, 480-482.

1609. Vantrappen, G., Hellemans, J. & Janssens, J. The pharmacological basis of the treatment of disorders of gastrointestinal motility. Acta Gastroenterol. Belg., 1967, 30, 646-658.

1610. Vantrappen, G., Texter, E. C., Jr., Babovka, C. J. & Vandenbroucke, J. The closing mechanism at the gastroesophageal junction. Am. J. Med., 1960, 28, 564-577.

1611. Vantrappen, G., Van Goidsenhoven, G., Van Derstappen, G. & Vandenbroucke, J. The value of electromanometric studies for the diagnosis and treatment of achalasia and hypertony of the cardia. Acta Chir. Belg., 1959, 2, 54-63.

1612. Varbanova, A. & Sokolov, V. Effect of mechanoreceptor stimulations of different intensities originating from the stomach and lungs on the afferent impulsation of the cervical vagus. Acta Physiol. Acad. Sci. Hung., 1967, 31, 107-113.

1613. Vass, C. C. N. Digestive system. Ann. Rev. Physiol., 1948, 10, 157-180.

1614. Vaughan, R. T. & Torek, P. Abdominal auscultation. Am. J. Surg., 1939, 45, 230.

1615. Veach, H. O. Studies on the innervation of smooth muscle. I. Vagus effects on the lower end of the esophagus, cardia and stomach of the cat, and the stomach and lung of the turtle in relation to Wedensky inhibition. Am. J. Physiol., 1925, 71, 229-264.

1616. Veach, H. O. The antagonistic action of morphine and atropine on the human stomach. J. Pharmacol. Exp. Therap., 1937, 61, 230-239.

1617. Venchikov, A. I. Bioelektricheskiye potentsialy zheludka. (Bioelectric potentials of the stomach). Mowcow, 1959.

1618. Vendel, S. Evacuation of the stomach in infants and prematures. Acta Physiol. Scand., 1946, 11, 380-385.

1619. Venturoli, L. Gastric evacuation in hypo- and hyperacidity states. Arch. Ital. Mal. Appar. Dig., 1968, 35, 364-374.

1620. Vesin, S. Biliary-digestive functional relations in the radiographic picture. I. Motoric response of the normal stomach to cholecystokinetic stimuli during normal biliary tract activity. Cesk. Radiol., 1967, 21, 152-163.

1621. Vesin, S. & Langemann, K. Problems of gastric motility following hyperventilation. Discussion on the paper by K. Langemann and K. Hofmann. Fortschr. Geb. Roentgenstr. Nuklearmed., 1973, 119, 245-247.

1622. Vierira, C. B., Meneghilli, U. G. & Simao, C. Gastrointestinal motility and the glycemic picture in the chronic stage of chagas' disease. Preliminary Observations, Hospital (Rio De J), 1967, 71, 1747-1755.

1623. Vishnevskii, A. A., Vilianskii, M. P., Livshits, A. V., Pavlychev, R. K. H., Kostiuchenko, V. I. & Kolpachkov, V. A. Control of the motor function of the gastrointestinal tract. Khirurgiia, (Mosk.), 1973, 49, 54-57.

1624. Visnovsky, P. Disorders of gastrointestinal motility in the irradiated organism. Cesk. Gastroenterol. Vyz., 1972, 26, 275-281.

1625. Visnovsky, P. The action of serotonin, lisuride and reserpine upon stomach-evacuation in rats. Cesk. Gastroenterol. Vyz., 1975, 29, 18-23.

1626. Vitebski, I. I. & Rudenko, V. I. Causes of disorders of the mechanism of evacuation from the gastric stump following its resection. Klin. Khir., 1974, 4, 22-27.

1627. Vizi, S. E., Bertaccini, G., Impicciatore, M. & Knoll, J. Evidence that acetylcholine released by gastrin and related polypeptides contributes to their effect on gastroinestinal motility. Gastroenterology, 1973, 64, 268-277.

1628. Vizir, A. D. & Grusha, V. N. Electrogastrographic changes in patients with peptic ulcer treated with vitaminized oxygenated beverage. Urach. Delo., 1968, 4, 19-21.

1629. Vlasov, P. V. X-ray pictures of the motor activity of the normal and pathologically changed stomach. Vestn. Rentgenol. Radiol., 1972, 47, 13-19.

1630. Volozin, F. A. Effect of anesthesia on stomach motor function in children in the immediate postoperative period. Vestn. Adad. Med. Nauk., 1972, 27, 51-55.

1631. Waddell, W. R. & Wang, C. C. The effect of vagotomy on gastric evacuation of high fat meals. J. Appl. Physiol., 1953, 5, 705-711.

1632. Wakabayashi, T., Yamanaka, T., Kunii, Y. & Ogo, T. Pharmacological effects of butozybenzyl hyoscyamine bromide on the motility and secretion of the gastrointestinal tract. Folia. Pharmacol. Jap., 1973, 69, 269-283.

1633. Walder, D. N. Muscularis mucosae of human stomach. J. Physiol., (Lond), 1953, 120, 365-372.

1634. Walker, G. D., Stewart, J. J. & Bass, P. The effect of parietal cell and truncal vagotomy on gastric and duodenal contractile activity of the unanesthetized dog. Ann. Surg., 1974, 179, 853-858.

1635. Walsh, A. Auscultation of the abdomen. J. Irish M. A., 1955, 36, 48.

1636. Walters, W., Priestly, J. T. & Belding, H. H., III. Vagotomy in the treatment of gastrojejunal ulceration; postoperative clinical and laboratory. J. Am. Med. Assoc., 1952, 148, 803-808.

1637. Wangensteen, O. H. & Carlson, H. A. Hunger sensations in a patient after total gastrectomy. Proc. Soc. Exp. Biol. Med., 1931, 28, 545-547.

1638. Wang, S. C. & Borison, H. L. New concept of organization of central emetic mechanism: Recent studies on sites of action of apomovphine, copper, sulfate and cardiac geycosides. Gastroenterology, 1952, 22, 1-12.

1639. Wang, S. C. & Chinn, H. I. Experimental motion sickness in dogs; functional importance of chemoceptive emetic trigger zone. Am. J. Physiol., 1954, 178, 111-116.

1640. Wang, S. C. & Glaviano, V. V. Locus of emetic action of morphine and hydergine in dogs. J. Pharmacol. Exp. Therap., 1954, 111, 329-334.

1641. Watson, B. W., Ross, B. & Kay, A. W. Telemetering from within the body using a pressure sensitive radio pill. Gut, 1962, 3, 181-186.

1642. Waugh, J. M. Effect of fat introduced into the jejunum by fistula on motility and emptying time of the stomach. Arch. Surg., 1936, 33, 451-466.

1643. Weber, J., Jr. & Kohatsu, S. Pacemaker localization and electrical conduction patterns in the canine stomach. Gastroenterology, 1970, 59, 717-726.

1644. Weber, J., Jr., Kohatsu, S. & Nelsen, T. S. Pacemaker localization and electrical conduction patterns in the canine stomach. Gastroenterology, 1969, 56, 1267.

1645. Weddle, C. O., II, Springfield, A. C. & Bass, P. The effect of parietal cell vagotomy and pyloroplasty on canine gastric emptying. Rev. Surg., 1974, 31, 285.

1646. Weddle, C. O., II, Springfield, A. C., Ormsbee, H. S., III & Bass, P. Parietal cell vagotomy and gastric emptying of liquids in the dog. Arch. Surg., 1974, 108, 83-86.

1647. Weisbrodt, N. W. Gastrointestinal motility. In E. D. Jacobson & L. L. Shanbour (Eds.), Gastrointestinal Physiology. Baltimore: University Park Press, 1974.

1648. Weisbrodt, N. W., Hug, C. C., Wiley, J. N. & Bass, P. Contractile activity of gastric antrum and taenia coli in the unanesthetized monkey. J. Appl. Physiol., 1971, 30, 276.

1649. Weisbrodt, N. W., Wiley, J. N., Overholt, B. F. & Bass, P. A relation between gastroduodenal muscle contractility and gastric emptying. Gut, 1969, 10, 543-548.

1650. Weitz, W. & Vollers, W. Studien uber Magenbewegungen. Ztschr. f. d. ges. exp. Med., 1925, 47, 42-69.

1651. Weitz, W. & Vollers, W. Ueber die Beeinflussung der Bewegungen des Magens und Darms durch Opium. Ztschr. f. d. ges. exp. Med., 1927, 54, 161-168.

1652. Wells, C., Tinckler, L. & Rawlinson, K. Postoperative gastrointestinal motility. Lancet, 1964, 1, 4-10.

1653. Welsch, K. H. Studies on stomach motility following surgery of the vagus nerve. Langenbecks Arch. Chir., 1968, 322, 708-712.

1654. Wenger, M. A., Clemens, T. L. & Cullen, T. D. Autonomic functions in patients with gastrointestinal and dermatological disorders. Psychosom. Med., 1962, 24, 267-273.

1655. Wenger, M. A., Engel, B. T. & Clemens, T. L. Initial results with the magnetometer method of recording stomach motility. Am. Psychol., 1955, 10, 452.

1656. Wenger, M. A., Engel, B. T., Clemens, T. L. & Cullen, T. D. Stomach motility in man as recorded by the magnetometer method. Gastroenterology, 1961, 41, 479-485.

1657. Wenger, M. A., Henderson, E. B. & Dinning, J. S. Magnetometer method for recording gastric motility. Science, 1957, 125, 990-991.

1658. Wertheimer, E. Inhibition reflexe du tonus et des mouvements de l'estomac. Arch. de physiol. norm. et path., Paris, 1892, 4, 379-35.

1659. Weschsler, R. L. & Roth, J. L. Measurement of rate of gastric emptying in man as determined by clearance of a radioactive colloid (AgI131) normal values, effect of urecholine chloride and morphine sulfate. Fed. Proc., 1954, 13, 161.

1660. Wheelon, H. & Thomas, J. E. Observations on the motility of the antrum and the relation of rhythmic activity of the pyloric sphincter to that of the antrum. J. Lab. Clin. Med., 1920, 6, 124-143.

1661. Wheelon, H. Observations on gastric and duodenal motility in duodenal obstruction. J. Am. Med. Assoc., 1921, 77, 1404-1406.

1662. Wheelon, H. & Thomas, J. E. Rhythmicity of the pyloric sphincter. Am. J. Physiol., 1921, 54, 460-473.

1663. White, E. H. Additional notes on gastrointestinal activity during avoidance behavior. Psych. Rep., 1964, 14, 343-347.

> Gastrointestinal activity in humans during avoidance responding was compared with that under conditions of passive reception of mild stimulation, as well as resting level. Results were contradictory to findings of Davis and Berry (1963) but tended to confirm those of Sternbach (1962) as well as to support Brady's (1963) work with peptic ulcer development in "Executive" monkeys.

1664. White, E. H. Surface recording of gastrointestinal motility. Psych. Rep., 1964, 14, 321-322.

> Several techniques for recording gastrointestinal activity are recounted. An input used in the electrical method is described in detail.

1665. White, E. H. Recording of vasomotor activity, skin resistance, and abdominal potential on the Bayamon Thoracopagus twins. Psych. Rep., 1969, 24, 391-398.

> Similarities and differences in the EGGs of thoracopagus twins is described. Electrodes were attached to the right side of the abdominal surface on one twin the left side of the other.

1666. Wilbur, B. G. & Kelly, K. A. Effect of proximal gastric, complete gastric, and truncal vagotomy in canine gastric electric activity, motility and emptying. Ann. Surg., 1973, 178, 295-303.

1667. Wilbur, B. G., Kelly, K. A. & Code, C. F. Effect of gastric fundectomy on canine gastric electrical and motor activity. Am. J. Physiol., 1974, 226, 1445-1449.

1668. Wilhelmj, C. M. Physiology of the digestive system. Ann. Rev. Physiol., 1952, 14, 177-204.

1669. Williams, K. R. & Walike, B. C. Effect of the temperature of tube feeding on gastric motility in monkeys. Nursing Research, 1975, 24, 4-9.

1670. Wilson, M. J., Dickson, W. H. & Singleton, A. C. Rate of evacuation of various foods from the normal stomach. Arch. Int. Med., 1929, 44, 787-796.

1671. Wolf, S. The relation of gastric function to nausea in man. J. Clin. Invest., 1943, 22, 877-882.

1672. Wolf, S. The Stomach. New York: Oxford University Press, 1965.

1673. Wolf, S. & Almy, T. P. Experimental observations on cardiospasm in man. Gastroenterology, 1949, 13, 401-421.

1674. Wolff, H. G. Emotions and gastric function. Science, 1943, 98, 481-484.

1675. Woodward, E. R., Robertson, C., Fried, W. & Shapiro, H. Further studies on the isolated gastric antrum. Gastroenterology, 1957, 32, 868-877.

1676. Woodward, E. R. & Schapiro, H. Threshold of enterogastric reflex. Proc. Soc. Exp. Biol. Med., 1956, 91, 509-510.

1677. Wormsley, K. G. Aspects of duodeno-gastric reflux in man. Gut, 1972, 13, 243-250.

1678. Wormsley, K. G. Response to duodenal acidification in man. IV. Effect on gastric emptying. Scand. J. Gastroenterol., 1972, 7, 631-637.

1679. Yamaguchi, T. Effects of 5HT on isolated strips of guinea-pig stomach. Br. J. Pharmacol., 1972, 44, 100-108.

1680. Yanagiya, I., Ohkubo, Y. & Shimada, M. Significance of longitudinal and circular muscle layer on the appearance of peristalsis. J. Physiol. Soc., Japan, 1958, 20, 462-468.

1681. Yaryura-Tobias, J. A., Hutcheson, J. S. & White, L. Relationship between stages of sleep and gastric motility. Behavioral Neuropsychiatry, 1970, 2, 22-24.

1682. Youmans, W. B. Innervation of the gastrointestinal tract. In C. F. Code (Ed.), Handbook of Physiology. Sect. 6, Alimentary Canal. Vol. IV. Motility. Washington, D.C.: American Physiological Society, 1968.

1683. Youmans, W. B. & Meek, W. J. Gastrointestinal inhibition in unanesthetized dogs during rectal stimulation. Am. J. Physiol., 1937, 120, 750-760.

1684. Young, G. C. The treatment of childhood encopressis by conditioned gastro-ileal reflex training. Behav. Res. Ther., 1973, 76, 12-15.

1685. Young, R. & Sun, D. C. H. Effect of glycopyrrolate on antral motility, gastric emptying and intestinal transit. Ann. N.Y. Acad. Sci., 1962, 99, 174-178.

1686. Yuh, L. On the innervation of the stomach of the Japanese frog. Jap. J. Med. Sci. Biophys., 1931, 2, 25-33.

1687. Zabolotny, I. L. S. Stomach motor function in acute dysentery (electrogastrographic data). Sov. Med., 1970, 33, 139-140.

1688. Zheleznyi, V. I. & Morozov, S. S. Motor function of the stomach following resection of its cardiac portion for cancer. Sov. Med., 1966, 29, 26-30.

1689. Zlatkina, A. R., Baklykova, N. M., Maisky, V. B. & Emel'Ianova, V. V. Characteristics of the motor function of the gastroduodenal portion of the digestive tract in peptic ulcer of the duodenum. Sov. Med., 1973, 36, 68-72.

AUTHOR INDEX

SUBJECT INDEX

SPECIES INDEX

About the Authors

ROBERT M. STERN, psychophysiologist and head of the Department of Psychology at The Pennsylvania State University, received the A.B. degree from Franklin & Marshall College, the M.S. degree from Tufts University, and, in 1963, the Ph.D. degree from Indiana University. His general research interest is the relationship between the autonomic nervous system and behavior. His specific interest is the effect of psychological factors on the functioning of the gastrointestinal system. Dr. Stern is the author of *Psychophysiological Recording,* coauthored with W. J. Ray and C. M. Davis (Oxford University Press, 1980); *Biofeedback,* written with W. J. Ray, (Dow Jones-Irwin, 1977); and numerous journal articles.

CHRISTOPHER M. DAVIS, associate professor of psychology at Simon Fraser University, Burnaby, British Columbia, Canada, received the A.B. degree from Indiana University and the Ph.D. degree from the University of Washington, specializing in psychophysiology. His research includes investigations into the control of autonomic and somatic muscular tensions. He is a coauthor of *Psychophysiological Recording,* along with R. M. Stern and W. J. Ray.